Reimagining the SCIENCE DEPARTMENT

WAYNE MELVILLE
DOUG JONES
TODD CAMPBELL

Arlington, Virginia

Claire Reinburg, Director
Wendy Rubin, Managing Editor
Andrew Cooke, Senior Editor
Amanda O'Brien, Associate Editor
Donna Yudkin, Book Acquisitions Coordinator

Art and Design

Will Thomas Jr., Director

Joe Butera, Senior Graphic Designer, cover and interior design

Printing and Production
Catherine Lorrain, Director

National Science Teachers Association
David L. Evans, Executive Director
David Beacom, Publisher

1840 Wilson Blvd., Arlington, VA 22201
www.nsta.org/store
For customer service inquiries, please call 800-277-5300.

18 17 16 15 4 3 2 1

Library of Congress Catologing-in Publication Data
Melville, Wayne, 1964-
Reimagining the science department / Wayne Melville, Doug Jones, Todd Campbell.
pages cm
Includes bibliographical references.
ISBN 978-1-938946-32-5 (print) -- ISBN 978-1-941316-79-5 (e-book) 1. Science teachers--Training of. 2. Science--Study and teaching--Research. I. Jones, Doug, 1957- II. Campbell, Todd, 1969- III. Title.
Q181.M425 2015
507.1--dc23

2015001252

Catologing-in Publication Data for the e-book are also available from the Library of Congress.
e-LCCN: 2015003567

CONTENTS

FOREWORD

Why would anyone want to write a book about something as universal as the secondary science department? Science departments are a common feature in secondary schools, and everybody knows their purpose—right? Most typically seen as convenient administrative units within the school, science departments have also been described as the engine room of the school, the place where the hard work of teaching and learning science occurs. More ominously, for school administrators they can also appear completely impervious to the most carefully laid plans for school improvement and reform. Put simply, the ubiquity of science departments means that they are often hidden in plain sight.

Even within the research literature, serious investigations into departments are relatively recent phenomena. In her seminal 1994 work, Leslie Siskin defined four aspects of subject departments that she believed were crucial to understanding their importance: (1) Departments are administrative units formed along their strong disciplinary boundaries; (2) they are the primary places for teachers' social interaction; (3) they have considerable power over what and how teachers teach; and (4) they judge what is considered acceptable in terms of teaching and learning for the discipline (Siskin 1994). These aspects have guided our work with departments, as both chairs or researchers, over a number of years.

However, two things have become obvious to us in undertaking our work. The first is that the functions—and nuances—of departments are still not well understood in the research literature, and even that limited understanding has made only a slow passage into schools. The second is that the critical role of the chair remains an area that is both understudied and undervalued. This situation is concerning, particularly when it is known that chairs are the linchpin between the principles and assumptions supporting proactive reforms and their successful implementation. The United Kingdom's Teacher Training Agency (TTA) phrases it this way:

> *A subject leader has responsibility for securing high standards of teaching and learning in their subject as well as playing a major role in the development of school policy and practice. Throughout their work, a subject leader ensures that practices improve the quality of education provided, meet the needs and aspirations of all pupils, and raise standards of achievement in the school. (TTA 1998, p. 4)*

FOREWORD

Science chairs are generally more experienced teachers with a solid grasp of both science content and pedagogy, and as middle managers they are in a unique position to influence the teaching and learning of both students and the teachers in their department. Yes, chairs have a responsibility to be good managers of the administrative side of the departments' operation. More importantly, they have the responsibility to be instructional leaders in their departments and to so help enact reforms to science education such as the *Next Generation Science Standards* (*NGSS*). However, if one looks at the history of reforms in science education, one sees a series of initiatives that looked good on paper but that stayed on the paper. Reformers often bemoan the inertia of teachers but continue to concentrate on the *what* of reform rather than the *how* of reform. The result is that increasingly cynical teachers see that the more change is called for, the more things stay the same. Clearly, such a situation does not benefit anyone, least of all the students in our classrooms. And make no mistake, students are voting with their feet and walking out of the discipline that we love. As Tytler (2007) points out, there is crisis in science education, characterized by secondary students developing increasingly negative attitudes toward science, a reduced participation in postcompulsory science education (especially in physics and chemistry), shortages of science-based workers, and a shortage of qualified science teachers.

This might all sound rather discouraging, but it also establishes the rationale for our work here. We firmly believe in the professionalism of science teachers, and as current, or past, chairs and science teachers, we understand and respect the pressures that act on both teachers and chairs. The purpose of this book is to assist science chairs, teachers, and administrators in beginning the task of reimagining the science department as a place where teachers are encouraged to question both their beliefs about science and the teaching and assessment strategies that develop in response to those beliefs. Only when teachers have the freedom and capacity to question their beliefs and develop their teaching and learning can real improvements in the teaching of the practices of science be sustained. This belief holds regardless of the school being urban, suburban, rural, public, or private. Between the three of us authors, we have taught in urban and rural independent schools in Australia, suburban public schools in Canada, and rural and urban Midwest schools—sometimes as the sole science teacher in the school. The writers of the vignettes, and our colleagues who have critiqued the earlier drafts, come from urban and rural areas in Ontario, New England, Georgia, and Texas. Different places and different teaching contexts, but for everyone who has contributed, the underlying departmental issues at the core of the work we are suggesting are the same.

The three-part structure of the book is designed to provide the reader with a firm foundation on which to base their actions. The first section, Chapters 1 and 2, places the science department in the context of its historical development, the relationship between the department and traditional science teaching, and the important (although under-recognized) role of the department in teacher professional learning. Most of us hold closely to an academic tradition of science education, and we need to recognize this before we can challenge it and its continuing impact on our teaching. The second section, Chapter 3, draws on the leadership and professional learning literature to consider the roles and responsibilities of science chairs in becoming instructional leaders. This section elaborates on many of the remarks in the National Science Teachers Association's position statements on leadership and professional development. We need to know the difficulties we will face if we are to move from recognizing, to challenging, to reforming our teaching and learning. To be prepared to reform means giving teachers good reasons to change. The pressure for reforms is not going to stop, so we need to be clear about the forces that drive that pressure and be proactive in dealing with them. The third section, Chapters 4 and 5, provides advice backed by research and experience on how to initiate reforms within the department and work with administrators to sustain and grow those changes over time. In this section we look at how chairs can make a start developing the credibility that is needed to influence the perceptions that departments have toward reforms, before finishing with the need to develop strong trusting relationships with school administrators in support of the work of the chair.

In our writing, we have constantly sought to avoid creating an "academic" book in the negative sense of two covers, pretentious prose, and wall-to-wall references. Such an approach does not reflect the day-to-day reality of departmental life. Conversely, where scholarly references add weight to the argument that we are making, we thought it appropriate that they be included. Theory and practice should not be seen as being diametrically opposed, they should inform and direct each other to improve the quality of teaching and learning. Our students deserve nothing less.

In each chapter we have included vignettes written by our colleagues that highlight the particular points made in the text; the issues that are faced are universal, and it is always nice to know you are not alone. We have also included questions to ask of yourself as a science teacher and as a chair. Such questions are important because to challenge the assumptions that underpin one's teaching, and then begin to really shift one's teaching and learning to a position that more closely resembles the ideals in reform documents such as the *NGSS*, is an intensely personal journey. Please feel free to rephrase the questions and use them in your own department as you see fit. As part of that journey, we would also like to invite you to send us anecdotes of your own trials, tribulations, growth, and successes connected to any of the chapters. If there is any way in which we can help you in your work, please don't hesitate to contact us.

Regards,
Wayne Melville, Doug Jones, and Todd Campbell
August 2014

ACKNOWLEDGMENTS

We have been fortunate to have worked with a number of talented and dedicated science teachers, educators, and administrators in the development of this book. To them, we wish to offer our sincerest thanks for their insights, comments, and criticisms. The work is stronger for their contributions.

- Anthony Bartley, Lakehead University, Thunder Bay, Ontario
- Wayne Bilbrough, Retired Chair, Lakehead District Schools, Thunder Bay, Ontario
- Ben Kirby, Jesuit College Preparatory School, Dallas, Texas
- Jeremy Peacock, EdD, Northeast Georgia RESA, Winterville, Georgia
- Jason Pilot, Lakehead District Schools, Thunder Bay, Ontario
- Matt Roy, Lakehead District Schools, Thunder Bay, Ontario
- Jeff Upton, Lakehead District Schools, Thunder Bay, Ontario
- David Welty, Fairhaven High School, Fairhaven, Massachusetts

We would also like to acknowledge the work of the manuscript reviewers who have kindly made a number of suggestions that have improved our work. Thanks.

ABOUT THE AUTHORS

Wayne Melville is an associate professor of science education at Lakehead University in Thunder Bay, Ontario. He taught secondary science in Australia from 1989 until 2005, and rose to become department chair. During his school teaching career, he completed a masters of science and a doctorate in science education and was a national finalist in a science teaching award organized by the Australian Academy of Science. Since moving to Lakehead University, he has published over 60 articles in the field of science education. His e-mail address is *wmelvill@lakeheadu.ca*.

Doug Jones has been a science chair with Lakehead District Schools in Ontario for 16 years and is an active member of both the National Science Teachers Association (NSTA) and the Science Teacher Association of Ontario (STAO). He is also the recipient of an STAO Service Award. Doug is working toward a masters degree and holds an Honours Specialist biology certification from the University of Toronto. His areas of expertise include mentoring preservice teachers and using scientific inquiry in secondary and elementary science classrooms. Doug's department has been the subject of chapters in two NSTA monographs on exemplary practice and has produced a video exemplar for the Ontario Ministry of Education on metacognition in science classrooms. He has also published articles in a number of journals. His e-mail address is *douglas_jones@lakeheadschools.ca*.

Todd Campbell is an associate professor of science education in the Department of Curriculum and Instruction at the University of Connecticut. He previously taught middle and high school science in Iowa. Recently, he collaborated extensively with teachers in funded professional development projects in Utah and Connecticut to develop and test curriculum in secondary science classrooms. He has served as guest editor for NSTA's *The Science Teacher* and has published articles in NSTA's *The Science Teacher*, *Science Scope*, and the *Journal of College Science Teaching*. His e-mail address is *todd.campbell@uconn.edu*.

1

A HISTORY OF THE SCIENCE DEPARTMENT

Departments are a ubiquitous feature of secondary schools; but where did they come from, what purposes do they serve, and what is the traditional role of the chair? These are not common questions, but it is necessary to ask them if we want to understand the importance of both the department and the chair in the teaching and learning of science. By knowing how the features we see in contemporary departments have evolved, we can begin to appreciate the power of departments in perpetuating a particular view of science education. If you understand this history, then as a chair (or aspiring chair) you will have a knowledge base from which to work in reforming science instruction in your department.

In this chapter, we will look at a number of interconnected themes that directly impact the work of the chair and his or her teachers. These themes include the curriculum traditions that have shaped science education, how the development of school science was closely tied to the professionalization of science, how school science departments were linked to science faculties in universities, why science departments occupy a privileged position within the hierarchy of school subjects, and how science departments exert tremendous administrative and social pressures on teachers to conform to a socially accepted view of "good" science teaching. To understand these themes, we need to step back to the beginnings of mass public education and look at the curriculum traditions that continue to influence our work as science teachers.

Curriculum Traditions

The development of mass education systems in 19th-century Britain and the United States involved powerful philosophical, political, and social forces arguing over the nature and purpose of education: what should be taught, who should teach, and who should be taught. From the 16th century onward, theories had developed that sought to explain why different societal classes should have different levels of education. One of the most widely held theories centered on the different "mentalities" that distinct classes possessed: The lower classes were supposedly only capable of simple, concrete thought, while only the upper classes were able to perform complex abstraction. Such rationalizations also served to legitimize the social order as it existed in societies on the edge of the Industrial Revolution. According to Hodson (1988) educational reform was driven by the political desire to maintain the status quo and was based on the assumption that the lower classes would only need the most rudimentary level of knowledge and that even this should be principally linked to their moral benefit. The maintenance of the social system required that "the lower orders … be *less knowledgeable* and have *less useful* knowledge" than the higher classes ([emphasis in original] Hodson 1988, p. 151). Education, where it was provided for the lower classes, consisted of reading and spelling and, to provide moral guidance, these subjects were taught from a religious perspective. The main educational institution for the majority of the working population was Sunday school, which was organized by churches. This meant that, starting in 1751 in Britain and 1790 in North America (and from there spreading across the British Empire), the Bible was the standard reading text. For the upper classes, education stressed the classics and religious instruction with the purpose of training and disciplining the mind, thus equipping the "Christian gentleman" for his position in life. The aim of disciplining the mind required high-status subjects that were closed to all other social classes. High-status knowledge was defined as being "not immediately useful in a vocation or occupation" (Eggleston 1977, p. 25). The importance of mental discipline was critical in shaping the curriculum offered by colleges across the United States during the 19th century and was also influential in defining Canadian educational policy. Kliebard (1986, p. 8) has argued that an "increased awareness of social transformation" had reduced the influence of mental discipline on education policy by the 1890s. This observation aligns with the growing influence of the Committee of Ten and challenges to the American ideal of the "common school."

The coming of the Age of Steam and the transformation of agrarian societies into urbanized, industrial societies led to efforts to reform education to provide for the expansion of commerce and industry. The differentiation of education that emerged in the 19th century continued to reflect the earlier political concerns of maintaining the social order while also beginning to reflect the new commercial and technological realities. Three traditions of curriculum began to coalesce around different perceptions of the nature and purpose of the school curriculum. Goodson (1993) describes these three traditions as the high-status academic tradition and the lower-status utilitarian and pedagogic traditions. These traditions are not fixed and have changed as perceptions around the value attached to different forms of knowledge have evolved. Broadly speaking, the academic tradition concerns itself with high-status content, which is presented in the theoretical and abstract and is readily evaluated through written examination. The academic tradition was seen as the preserve of the ablest students who were destined for the learned professions and high-level administrative or commercial positions. The utilitarian tradition was principally concerned with more practical forms of knowledge that are not as readily evaluated by written examination. Practical forms of knowledge, including numeracy, literacy, and commercial and technical education, were seen as appropriate for those destined to nonprofessional occupations. The pedagogic tradition is one in which education is not seen as preparatory for career paths; rather, it is concerned with how children learn and emphasizes inquiry, discovery, and the use of activity-based teaching strategies.

Understanding the context of educational change in the 19th century is important because it connects the evolution of science education to contemporary school science, a subject firmly entrenched in the academic tradition despite nearly 50 years of reform efforts that echo the pedagogic tradition. The subject we see today, the ways in which it is predominantly taught, and the privileged position that it holds in schools is very much the product of a persistent push to raise the status of science. It was this drive for high status that has shaped both science education and the appearance of the first science departments at the turn of the 20th century. These developments continue to impact the life and work of contemporary science departments and chairs.

Science and School Science Education

The development of science as a school subject cannot be readily grasped without also understanding the development of science as we know it, which happened in the first half of the 19th century. During this time, natural philosophy in the tradition of Galileo transformed into what we now recognize as science. In 1831, the British Association for the Advancement of Science (BAAS) was formed with the intention of professionalizing science, and it institutionalized many of the features we now take for granted. A driving force behind BAAS was William Whewell (1794–1866), who was active in promoting the superiority of the "pure sciences" over the technological applications of science—a status divide that continues to this day. A key strategy was to increase the status of the natural philosophers working in universities. To this end, Whewell coined the term *scientist* in 1834. A decision was made in the early days of BAAS to partition the annual meeting into the disciplines, a decision that still shapes the structure of contemporary science. Both Galilean pure sciences of physics and chemistry were included, as were the "applied" sciences of botany, zoology, and geology. Biology was not included because it was considered a practical pursuit, valuable only when it "contributed to useful ends, such as the extermination of insects destructive to timber in the dockyards" (Layton 1973, p. 21). The status of biology only increased as its utilitarian potential developed, and it began to claim disciplinary rigor through the efforts of scientists such as Pasteur and Koch. It was not viewed as a hard, empirical science until the 1960s, when molecular biology gave the discipline a unifying, quantifiable basis (Goodson 1993). The success of BAAS in raising the status of both science and the scientist served as a model for the 1848 formation of the American Association for the Advancement of Science (AAAS).

With their new disciplinary structures, the newly professionalized university-based scientists set about boosting the status of scientific knowledge and the resources available to them as scientists by appropriating both disciplinary content and access to the disciplines. As we have seen, a discipline that seeks to possess high-status must concern itself with content that is presented in the theoretical and abstract and is readily evaluated through written examination. Layton (1973) sums up how this drive for high status was achieved:

> *A necessary, although by no means sufficient, requirement for this remarkable surge of intellectual activity was the introduction into scientific thought of conceptions of which there were no directly observable instances. ... The application of mathematics to the description of nature was a further vital step in the growth of modern science. (p. 168)*

In addition to this abstraction and narrowing of the scientific discourse, science became divorced from any technological application: "Advocates of scientific education took care to distinguish it from technical instruction, and to emphasize its liberal and academic value rather than its industrial and commercial utility" (McCulloch, Jenkins, and Layton, cited in Hodson 1988, p. 145). Remember, for knowledge to have high status, it could not be "immediately useful in a vocation or occupation" (Eggleston 1977, p. 25). An important link between this effort to raise the status of science and science education was the 1839 establishment of the German chemist Liebig's laboratory at the University of Giessen. Brock (1975) has identified this date as the starting point of modern science education, as Liebig's move established the ideal of the research school in which research and inquiry were pursued as ends to themselves. Hodson (1988) writes that Liebig's laboratory "provided the concept of 'pure laboratory science' which was to dominate school science curricula, especially for the more able child, throughout modern times" (p. 141) . In pushing to raise its status, pure laboratory science allowed science to become abstracted and separated from the everyday experience of the ordinary person. This separation is exemplified in the case of natural history, which shifted from a "study of nature not books" utilitarian pedagogy to the disciplines of botany and zoology with their "technical and required laboratory work that was exacting and precise" (Sheppard and Robbins 2007, p. 201).

The power of the newly professionalized science in shaping the purposes of science education is illustrated by the fate of the ideas of the Reverend Richard Dawes. Dawes was an influential early science educator who believed that science education should help all students to observe, reflect, and reason. The science curriculum, according to Dawes, should encourage students to consider science in the context of their everyday life and from there move toward explanations of "a philosophic kind" (Layton 1973, p. 41). As Dawes stated:

> *In subjects of this kind, and to children, mere verbal explanations ... are of no use whatever; but when practically illustrated ... by experiments, they become not only one of the most pleasing sources of instruction, but absolutely one of the most useful. (Report of the Committee of Council on Education for 1847–8, vol I, 12–18 cited in Layton 1973, p. 42)*

To this end, he instructed the teachers in his schools in both content and the utilitarian value of science as it applied to life. His most influential book, *Suggestive Hints Towards Improved Secular Instruction,* was first published in 1847 and ran to six editions. It became the prescribed text for teacher education, with the Irish National Board presenting a copy to all trainee teachers (Layton 1973). In stark contrast to the academic tradition, the heart of Dawes's work was a belief that scientific knowledge should "work in a manner which engaged the intelligence of children of the labouring poor, of whom it had previously been taken for granted that they should have 'no business with anything where the mind is concerned'" (Layton 1973, p. 43).

However, Dawes's ideas and practices were no match for the growing status and power of the newly professionalizing science, and over time the influence of Dawes receded. Diminished, but not forgotten, his pioneering work can still be seen as the foundation on which subsequent reform efforts have been built. General science in the early 20th century, science as inquiry, and the *Next Generation Science Standards* (*NGSS;* NGSS Lead States 2013) all share the same commitment to experience as a strategy for developing explanations and the need for science teachers to be skilled in both science and pedagogy. Importantly, the work of Dawes was also the precursor to our contemporary understanding of the role of the chair as both curriculum expert *and* instructional leader.

The high status that science had achieved by the middle of the 19th century had important implications for the resourcing of schools. Achieving high status brought with it the capacity to acquire and retain resources and also limit access to the new scientific profession: Science was "linked to the university elite, a largely upper class, masculine elite, rendered in the image and language of that elite and perfect for sponsoring the interests of that elite" (Goodson and Marsh 1996, p. 73). In terms of resources, pure laboratory science in Britain was being heavily promoted and supported by the national government in the 1850s. The first high school laboratories in the United States began to appear in the 1870s: Massachusetts' Dorchester High School in 1870 and Gives High and Normal School in 1871. Canadian schools began to develop laboratories in the 1890s, in response to the influence that university science exerted on the developing public education systems:

> *By 1890 science had gained acceptance. ... Ontario high schools were now required to have laboratories in order to gain collegiate status. The "pure" science of the university disciplines meant, as it would continue to, that applied aspects received short shift in the curriculum. (Tomkins 1986, p. 87)*

There remained, however, powerful opponents to the direction that science education was taking this time. In 1862, Michael Faraday, who always referred to himself as a natural philosopher—never a scientist—presented evidence to the British Parliamentary Public Schools Commission. Speaking against the manner in which science was being taught in Britain, Faraday argued that "if you teach scientific knowledge without honouring scientific knowledge as it is applied, and those who are there to convey it, you do more harm than good. You only discredit both the study and the parties concerned in it" (Public Schools Commission 1864, p. 380). Faraday's evidence, along with the evidence of Charles Lyell, Richard Owen, and Joseph Dalton Hooker was ignored, and in 1867 the academic tradition of science education asserted its preeminence, a position it retains to this day.

In that year, the BAAS produced the highly influential *Scientific Education in Schools* report. This report shaped a science curriculum that stressed the values of the BAAS and entrenched the power and prestige of the upper classes. The report promoted an elitist professional training of future scientists in the pure sciences through mental training (Layton 1981). Universities began to control access to science through entrance examinations, which gave the universities immense powers of accreditation, reinforcing the significant influence that universities and professional scientific communities held over the teaching and learning of school science. For science education, the report formally established the superiority of disciplinary scientific knowledge: "education reformers ... produced a science curriculum that marginalized practical utility and eschewed utilitarian issues and values related to everyday life ... [reflecting] the BAAS's newly achieved divide between science and technology" (Aikenhead 2006, p. 13). These purposes of science education also came to dominate in the United States and Canada. While the report of the Committee of Ten from 1893 is seen as being particularly influential in shaping the general school curriculum, evidence shows that the impact in science education was more influenced by the differentiation in college entrance subjects and the related issues of the nature of the pure sciences and their capacity to "exercise the mind." The ideal of the common school began to fade by the end of the 19th century as demands were made for different curricula for different postschool occupations (Goodson and Marsh 1996). In Canada, the leading architect of public education in Ontario, Egerton Ryerson, promoted a differentiated curriculum and school system: collegiate schools (with science laboratories) for university preparation and secondary schools and vocational institutes (without science laboratories) for the everyday life curriculum. These school titles remain in place today, even as the differentiated curriculum is less explicit.

The differentiation of the curriculum and the value placed on high-status disciplines had an important impact on the organization of schools. Subject departments, often closely aligned with university disciplines, formally began to appear in the mid-19th century. This evolution of science as a school subject and the changing role of the science teacher are the foci of the next section.

The Department: Subject and Teachers

Science as a School Subject

From medieval times until the 19th century, teaching was far from standardized, with each school developing its own idiosyncratic strategies and organization. There was no "idea of establishing a hierarchy of ability or a sequence of learning" (Reid 1985, p. 296). The rise of public education systems and the importance attached to university entrance examinations by the increasingly high-status science disciplines replaced the previous laissez-faire system. The demands of the university entrance examinations drove the content that schools taught, how that content was taught, and the evaluation of that learning. To accommodate the demands being placed on them by university control of content and accreditation, schools adopted standardized systems of timetables, lessons, and school subjects. This development began in Britain in the 1850s and was solidified in 1917 with the establishment of the school certificate, which focused on the definition of the content knowledge and evaluation of examined knowledge in university preparatory subjects. These academic subjects, including science, soon gained the ascendency in the life of the school, a reflection of their high status. The rise of the academic subject was replicated in the United States, where the 1893 report of the Committee of Ten clearly elevated the status of the academic subjects to align with the admissions requirements of the university disciplines. John Dewey quickly recognized the differentiation in education that was entailed and railed against it. Those students who did not fit the academic tradition would be left with the utilitarian tradition, the course of their life determined by their schooling. As Dewey stated, such an education was "an instrument in accomplishing the feudal dogma of social predestination" (Dewey 1916, p. 318).

One response to the academic tradition and increasing pressure for specialization of school science was the development of general science courses in both the United States and Britain. A consequence of the massive increase in the secondary population in the early part of the 20th century was a relative decline in enrollments for academic science courses. Among the reasons put forward for this decline was the increasing distance between specialized scientific research and the general population. As a commentary in *The Nation* in 1906 noted, "science has withdrawn into realms that are hardly understanded [sic] of the people. ... To-day, no layman may fairly hope to keep up, and all sorts of popularization meet with increasing difficulty" (*The Nation* cited in Rudolph 2005, p. 363). General science, which was not tied to the disciplines, introduced topics that were more aligned to the interests and needs of the newly urbanized population such as public health, heating, and refrigeration. General science was not the work of university scientists, it was the work of professional science teachers who had embraced the problem-solving conceptualization of science education promoted by Dewey and other early psychologists. This did not mean that general science shunned disciplinary science, rather the view was that students would learn how

> *scientific methods could be used to solve the many problems they encountered in their communities or about the scientific principles underlying the operation of the assorted appliances of the time, all the while working toward an understanding of the value of more conceptually sophisticated disciplinary knowledge. (Rudolph 2005, p. 381)*

Teachers of general science also sought to promote a broad vision of science to the public, one that was more relevant to the population and also capitalized on the intense public interest in the technological and industrial marvels of the age. A major teaching strategy that developed was the problem approach, in which students were given a concrete problem to solve in class or at home. Over time, this approach came to have more of an engineering focus (a historical precedent for the *NGSS*?), thus moving away from Dewey's problem solving, which stressed the application of the scientific method to students' everyday questions.

By the late 1930s, general science was well established across the United States and its territories as well as the nations and colonies of the British Empire. Well established does not mean, however, that general science was accorded the same prestige as science in the academic tradition. General science was scorned by powerful critics, who continued to push a highly specialized vision of school science that prepared future professional scientists. The major criticism centered on the supposed detachment of the processes of science from the conceptual knowledge base of the disciplines. For many scientists, this perceived downgrading of theoretical knowledge was a misrepresentation of science, and it threatened their privileged position. Along with the huge surge in scientific investment during and immediately after the Second World War, research scientists sought to reinforce the disciplinary structure on school science as a way of reshaping the public's perception of science and scientific research and reinforcing both the importance and status of science. The judicious application of discipline-based science was seen as offering a way to solve apparently intractable problems (e.g., the Green Revolution) or, literally, reach for the stars.

In 1962, at the height of the Cold War, Schwab published *The Teaching of Science as Enquiry*. This work opened a new round of introspection into the meaningful purposes of science education, again questioning the now traditional role of school science as preparation for students to become scientists. Hodson (2003, 2011) has summarized a number of common threads in these conversations: the meanings of scientific literacy; the relationships between science, technology, society, and the environment; the economic imperatives driving science education; the nature of science; and how to improve the teaching and learning of science. However, over time something of a disconnect developed between the continued academic tradition of school science and these broader discussions. Within education policy documents, science is increasingly being understood as less objective, mechanistic, and decontextualized than it had been stereotyped to be. Science is now increasingly being portrayed and understood as subjective, tentative, deeply contextualized, local, and reliant on human inference, creativity, and imagination. Science is also seen as being able to make an important contribution to solving national and international problems. The *NGSS* state that

> *although the intrinsic beauty of science and a fascination with how the world works have driven exploration and discovery for centuries, many of the challenges that face humanity now and in the future—related, for example, to the environment, energy, and health—require social, political, and economic solutions that must be informed deeply by knowledge of the underlying science and engineering.* (p. 7)

The apparent disconnect between the reform documents and school science occurs across schools and university faculties of science and education. As the

European Commission's High Level Group on Human Resources for Science and Technology reported:

> *Unfortunately, science education has been inclined to isolate itself from the rest of education and has tended to be separated by society into its own subculture. There is a strong tendency to regard the teaching of science not as an area of educational development of the student, but solely for the pursuit of the subject matter. Science education is viewed as the learning of 'science knowledge', rather than 'education through a context of science'. There is thus pronounced confusion between science on the one hand and science education (that which is promoted in schools) on the other.* (2004, p. 9)

We want to be very clear here: *There is nothing to be achieved by seeking to blame teachers for not implementing the lofty ideals of reform documents.* Contemporary science teaching has been shaped into the academic tradition for nearly 200 years by powerful historical, political, social, economic, and educative forces. These forces have socialized teachers into a particular set of beliefs about what good science teaching looks like—a focus on content knowledge that is readily examinable. For an individual teacher to challenge these beliefs is problematic because it means challenging both his or her own beliefs *and* the instructional strategies within a department where such beliefs and instructional strategies are generally taken as given. One core reason for writing this book is to help department chairs *understand* how the beliefs and instructional strategies that they see in themselves and their departments can evolve (see Vignette 1, p. 18). The second core reason for this book is to provide some guidance in how to reimagine the department as a place in which teachers are prepared to work together to improve the teaching and learning of science. The absolute need to work together to effect meaningful change leads us to outline the development of the contemporary science department and how it has come to wield enormous power, both within schools and over its teachers, while acting simultaneously as both a community and an organization.

The Role of the Science Teacher

With some noticeable exceptions, the role of the science teacher since the 1850s has been largely dictated by the academic tradition of science education. Committed to a body of abstract knowledge, and examinable in line with university admissions requirements, school science in the second half of the 19th century was taught by generalist scientists: those who could teach science "in such a way and to such a standard as will ensure success in the ... examination" (Goodson 1993,

p. 30). These scientists formed the first science departments, with the first modern usage of the term *department* by Kilpatrick in 1905. He argued that students in both elementary and secondary schools would benefit from being taught by the same teacher for several years. Financially there were benefits as well. Teachers would seek out and buy specialist equipment that could be used over longer periods of time. Kilpatrick's ideas were not taken up in elementary schools but were adopted in secondary schools, especially in Britain. Bureaucrats charged with developing mass schooling in the early 20th century saw the departmental structure as aligned with their concerns for "efficiency, rational management, and differentiation" (Siskin 1994, p. 44). The situation was different in the United States, where subject departments were slower to align with their disciplinary counterparts in the universities, and factors such as school size and function were more important.

In the early part of the 20th century, there was a growing recognition by educational theorists that the role of the science teacher was different from the research scientist working in a university. The role of the science teacher, as exemplified in the development and practice of general science, came to be understood as promoting rational scientific thought and an appreciation of science to the general public. At the same time, increasing university specialization in disciplinary science was an ongoing issue for science teachers, providing a constant reminder of the academic tradition and the power of universities over the school curriculum. By the 1920s, secondary teachers were being educated as specialists, a development that cemented the strong bonds between the discipline and the department. Some saw this as a retrograde step:

> *A generation ago there were men who were scientists without being scientific specialists. ... As an instrument of education that Victorian type of scientist held an immeasurable advantage over the botanist or biologist or chemist of to-day, whose attention is too often narrowed to one sub-division of one sub-branch of science. (Fyfe 1934, p. 658)*

The trend in teacher education toward specialization in disciplinary science was overwhelming, and by the 1930s departments staffed by specialists had become the dominant form of school organization, representing a "developed, established, and legitimated system of matching the organizational divisions of departments to the knowledge divisions of subjects" (Siskin 1994, p. 56). To this day, there is evidence of a strong preference by department chairs for incoming science teachers to be subject specialists, initially educated in faculties of science before gaining teaching qualifications (Harris, Jensz, and Baldwin 2005). The main reason for this preference is the belief that science faculties provide beginning teachers with the "necessary 'depth' in content knowledge" (Harris, Jensz, and Baldwin 2005, p. 22). This enduring link between high-status science as the discipline and science as the subject is an important feature in understanding the culture and power of science departments within schools. Indeed, some critics have questioned science itself as a culture of power (Calabrese Barton and Yang 2000). And so we arrive at the culminating part of this chapter, an understanding of the particular qualities that permit the department to shape, and be shaped by, the work of teachers.

Departments as Communities and Organizations

Departments are not just convenient administrative structures within secondary schools, although that is often how they appear. As we have seen, departments developed in response to the changes that occurred in both science (as both discipline and subject) and the increasing importance of public education. As a result of these changes, it has become clear that departments actually have two main functions that influence the lives and work of teachers. Contemporary science departments are simultaneously learning communities, which powerfully influence "what and how teachers teach" (Siskin 1994, p. 5) and administrative organizations within secondary schools. Nowadays, words such as *community* and *organization* are often used rather carelessly in conversations about education, so we need to be very clear what the words mean in the context of a chair looking to lead the department. If chairs are to be effective in their work, they need to understand that departments are both communities *and* organizations, as the perception the chair holds will produce very different leadership actions. A perception of the department as a community will lead a chair to rely on strategies aimed at changing the culture of the department. This requires more time but will be more effective in making lasting changes to teaching and learning. On the other hand, a perception of the department as an organization will lead to chairs relying on bureaucratic solutions to the issues that arise in their work. These forces are efficient in changing departmental structures over the short-term but do not promote fundamental changes to teaching and learning. A chair that sees the department as both organization and community is in the best position to judge the most appropriate approach to the issues being faced. There are times when bureaucratic approach is most appropriate, and there are times when a long-term cultural approach is needed. And, of course, there are also times when there is no obviously correct approach.

Communities

One way to understand the idea of the department as a community is to see a group of teachers who share a common sense of identity, a common sense of what it means to be a science teacher, and a relatively common set of instructional strategies. As science teachers we identify ourselves as teachers *of* science. Our educational success in the sciences gives us a sense of purpose as to what is important in our subject, how it should be taught, and why it should be taught. The discipline of science is important to us because it provides a sense of direction and values to our teaching and learning. A science teacher's identity is a powerful determinant of his or her view of the world and place in it, and the meanings and instructional strategies that he or she develops toward the teaching and learning of science.

The meanings that teachers attach to their work develop whenever teachers discuss issues of subject matter, instructional strategies, and the values that are important to them in the context of their work. Meanings are important in that they shape our teaching and instructional strategies. By far the majority of secondary science teachers have been educated in the academic tradition of science and, as a consequence, have a relatively consistent view of what is important in science classrooms and how they should go about their work:

> *This culture is strongly represented in school science discursive practices, supported by resources such as textbooks, laboratories and their associated equipment, timetabling arrangements and by assessment and reporting traditions. Another aspect is the force of long habit of teachers who have developed effective ways of delivering canonical content, who may lack the knowledge, skills and perspectives required for the effective teaching of a different version of school science. (Tytler 2007, p. 18)*

This history of increasing subject specialization within the academic tradition has serious pedagogical implications for science teachers. The historical preference for science teachers to be subject specialists has tended to restrict the range of answers that are acceptable when we start to ask questions about "what science is, who does science, what belongs in the science curriculum, and how best to 'deliver' the content" (Carlone 2003, p. 308). To begin asking questions such as these is to challenge the academic tradition of science education, and that is a difficult undertaking in any well-entrenched community. However, that does not mean, however, that these questions should not be asked.

Chairs need to recognize that communities derive their power from the nature of the work that their teachers undertake. Science teachers, rightly, tend to believe that only science teachers possess the necessary skills, knowledge, and academic orientations necessary to judge the quality of their own work. For most science teachers, community membership within a department is understood to involve a solid grounding in the pedagogy and the discipline of science in the academic tradition. Communities, however, can change; science teachers "do not always share the same views about what constitutes good teaching" or the same views on the "nature of science itself" (Wildy and Wallace 2004, p. 100). Within our departments there are teachers who will see science as "a process of personal sense making that helps people survive in their environment," even as the dominant traditional view will hold that science is "a set of universal truths that describe the operation of the natural world" and the purpose of science education "is to deliver that knowledge" (Wildy and Wallace 2004, p. 109). Following on from these two descriptions of science come two coexisting conceptions of work in the science department that chairs need to acknowledge and work with. Under the first conception, the science department operates as a community, whereby "the goals of inquiry, individuality and freedom sought for students in the classroom are also sought for teachers in the department" (Wildy and Wallace 2004, p. 109). Under the second conception, the science department operates as "a tightly organized and orderly place to work [where] there is one best way of teaching and a single best way of assessing students' learning" (Wildy and Wallace 2004, p. 109). This second conception emphasizes the organizational side of the department and the power that it has to shape the work of science teachers.

Organizations

Organizationally, departments are the administrative units into which secondary schools are divided. As administrative units, departments have been described as the engine rooms of secondary schools, possessing a unique position of power within the school, particularly through their capacity to influence *what* teachers teach and *how* they teach it. In their discussion of the politics of schools, Blenkin, Edwards, and Kelly (1997) describe the politics of departments as being concerned with the "control of territory, the distribution of resources, the acquisition of status, and participation in the decision-making process" (p. 222) . Viewed in this way, departments have powerful tools available to socialize teachers into a commonly accepted sense of what good science teaching looks like. While membership of the department provides a place for social interactions, support, and the shaping of a teacher's identity, it is also "a formally sanctioned administrative unit, [with] the authority to command and dispense far more tangible rewards and sanctions" (Siskin 1994, p. 114). These tangibles can include class allocations, resource allocations, access to professional development activities, and course offerings. As a chair, the control of these tangibles is a real source of power, albeit one that must be used wisely. How often have we seen seniority used as the sole basis for selecting who goes to a conference? How many times has the least experienced teacher been left with the classes that everyone else has passed over? As organizations, departments have a tendency toward conformity and the ability to reward those who conform. This makes departments "a most significant organizational and political division within the secondary school" (Ball cited in Blenkin, Edwards, and Kelly 1997, p. 222). Chairs, therefore, must understand that part of the role is intensely political. The question is, how do we as chairs balance the department as community with the department as organization?

A Balancing Act

As teachers of a historically high-status subject, science teachers' professional identity is tightly bound up in their subject. As communities, departments shape a common sense of identity, a common sense of what it means to be a science teacher, and a relatively common set of instructional practices. As organizations, departments have a great deal of political power to shape the teaching and learning of science through the access that they offer to resources. This political power is amplified within schools because science, with its rich academic tradition, is the source of great privilege and power within a school. The academic tradition sees science as a difficult subject, a subject only suitable for the most able students. If a subject traditionally believes that it caters to the most capable students, then it holds political power within a school when decisions regarding resources, such as equipment, teaching pedagogies, time, and teaching assignments, are being made. It is not difficult to see why science (and mathematics, another high-status subject) teachers often see themselves, and are seen by members of the public, as occupying the apex of the subject hierarchy. The danger of this is that any reforms that seem to threaten this position are resisted. That resistance may be active or passive but will most likely be in the form of adopting the language of the reforms to describe the current work of the department instead of making actual changes.

As a chair, or if you have ambitions to exercise leadership within your department, it is important to understand the powerful forces that will influence both your own thoughts and actions but also the thoughts and actions of your colleagues. Understanding these forces is crucial if you are looking to lead a department that is planning to align itself with reform documents such as the *NGSS* and working to integrate the practices of science into its teaching and learning. To challenge those forces requires power: the "ability to cause or prevent change" (Bybee 1993, p. 157). To begin the task of reimagining the science department as a place where teachers are encouraged to question both their beliefs about science and the instructional practices that develop in response to those beliefs is to challenge deeply held positions. Chairs need to work actively with their colleagues while remaining aware that the role of chair is an intensely "political one because it involves the creating, organizing, managing, monitoring and resolving of value conflicts, where values are defined as concepts of the desirable ... and power is used to implement some values rather than others" (Brundrett and Terrell 2004, p. 17).

To promote values that support reimagining the department toward the reform of teaching and learning is a political decision that must take into account the dual nature of the department and its relation to science education. Our intention in this chapter was to provide you with an understanding of where the department has come from, the tensions that exist within science education, and how the department possesses characteristics that can help—or hinder—reform. With these thoughts in mind, the next chapter will explore in more detail the issues of school science as currently taught; the main thrust of the various reform efforts of the past half century; and why the department, as both community and organization, should be considered the principal site of reform efforts. Each of these issues has great implications for the role of the chair.

Summary

- A timeline of historical events that shaped science teaching, chairs, and departments is shown in Figure 1.1.
- School science is firmly grounded in an academic tradition that values abstract knowledge that is easily evaluated.
- The academic tradition is closely aligned with the high-status and power that science accrued in a process that began in the 1830s.
- Science departments formed in response to the content and entrance demands placed on schools by universities.
- Science teachers are socialized into the academic tradition, with its associated identities, meanings, and instructional strategies. As a consequence, teachers do not readily challenge contemporary science teaching.
- As teachers of a high-status subject, science departments have considerable political power within schools.
- Departments control what is taught, and how it is taught, through their dual roles as organizations and communities.
- Chairs have political power, but must understand that change can only start from where their individual and colleagues' practice is currently located.

Figure 1.1

TIMELINE OF HISTORICAL EVENTS THAT SHAPED SCIENCE TEACHING, CHAIRS, AND DEPARTMENTS

- In **1751** in Britain and **1790** in North America the Bible was the standard reading text.
- In **1831** the British Association for the Advancement of Science (BAAS) was formed with the intention of professionalizing science
- Whewell coined the term *scientist* in **1834**.
- The success of the BAAS acted as a model for the **1848** formation of the American Association for the Advancement of Science (AAAS).
- Reverend Richard Dawes published *Suggestive Hints Toward Improved Secular Instruction* in **1847**, which ran to six editions.
- Pure laboratory science in Britain was heavily promoted and supported by the national government in the **1850s.**
- In **1862** Michael Faraday spoke against the manner in which science was being taught in Britain.
- In **1867** the academic tradition of science education asserted its preeminence.
- The first high school laboratories were established in the United States at Massachusetts' Dorchester High School in **1870.**
- Canadian schools began to develop laboratories in the **1890s.**
- In **1893** the Committee of Ten in the United States elevated the status of the academic subjects through alignment with the admissions requirements of the university disciplines.
- In **1962** Schwab published *The Teaching of Science as Enquiry*.
- The *National Science Education Standards* were published in **1996**.
- In **2013** the *Next Generation Science Standards* were published.

Vignette 1

BEN KIRBY

Ben Kirby is a science educator and administrator at the Jesuit College Preparatory School of Dallas. He earned an undergraduate degree in biology and a masters degree in higher education administration from the University of Kansas, Lawrence. He is currently working on a doctorate in curriculum and instruction from the University of North Texas. We asked him to write about the evolution of both his own teaching and his department. As you read Ben's story, think about how your own teaching and the work of your department has changed over time.

Teaching science and being an active member of a secondary science department is intellectually rewarding and philosophically satisfying. To that end, I am passionate about my vocation as a science educator because of the structured and spontaneous experiences it provides me and others to discuss perceptions and understandings of the natural world.

My approaches to preparation, teaching, and evaluation have evolved since I started teaching in 2006 from working as an individual entity in determining the "best" curricular and pedagogical methods to fully embracing the synergistic structure provided by my science department and professional organizations (e.g., National Science Teachers Association [NSTA] and the Association for Science Teacher Education [ASTE]). In reflecting about why I teach the way I do and what led me away from the autonomous way of working, I found myself focusing on my secondary and postsecondary science courses. My former teachers and classrooms quickly came to mind, and I realized that all of my teachers had been educated before the effects of the Russian *Sputnik* launch fully trickled into education and solidified the inquiry approach to teaching and learning that is found in current research and teacher education program curricula. Moreover, every one of them implemented a pedagogical approach that conveyed content and developed skills almost entirely through teacher-driven lectures and recipe-like labs. Despite my own thoughts and teacher education experiences, my first-year classroom and preparation strategies incorporated this model and, therefore, fostered explicit and implicit delineations between the teacher and student roles in knowledge acquisition. There were few opportunities for students to "discover" information through inquiry experiences because I was simply playing the role of knowledge deliverer. As a result of graduate studies, diverse mentorship interactions, and my interaction with students, I have since morphed into a teacher who tries to affirm, or modify, the cognitive framework of my students by *intentionally* incorporating new scientific understandings, diverse and innovative pedagogical approaches, and the ever-changing degree of students' prior knowledge. Attending professional development conferences and reading about various innovations in pedagogy (e.g., flipped classroom, inquiry learning, project-based learning) have supported my goal of continuing to move farther away from the lecture model and constantly incorporate new, or refined, experiential opportunities for students that safely address misconceptions, build accurate conceptions, and apply learning to higher-level thinking actions.

While I rely heavily on outside resources to help hone my knowledge and abilities, a major resource for learning is the teachers in my department. My department is composed almost entirely of teachers who identify as individuals who specialize in the content and skills necessary to prepare students to be critical thinkers. Specifically, we explicitly try to develop individuals who are both appreciative of natural law and aware of interdependent environmental relationships. To maintain and support this identity, we collaborate on an internal department basis and are constantly evaluating how to use standards (e.g., *NGSS*) in our preparations and implementations. The *NGSS* have provided us with an interesting standardization, or framework, on which we can infuse our culture and teacher personalities. They are also an excellent and reliable touchstone for discussions that involve new scientific discovery and curriculum. Although very few teachers in my department have published research or engaged in "high-caliber" research projects, we as a department seem to have a consistent understanding of what good science is because of the communications from professional organizations (e.g., *The Science Teacher*, published by NSTA, and *The Journal of Science Teacher Education* that shares research from ASTE members) and conference sessions. In addition to meetings during the academic year, we have end-of-year reviews of the curriculum to determine what changes are necessary as they relate to the integration of new content or alterations in methods of instruction.

One of the greatest challenges in maintaining a collaborative unit and getting traction with the review process is ensuring everyone is open to change and willing to engage in constructive criticism of self and others. The teachers of my department are very passionate about their vocation and span the spectrum regarding how receptive they are to change; several teachers are willing to change direction when they understand the rationale for the change while others become emotionally charged when confronted with a recommendation or question. This diversity challenges the chair leadership position because it requires a constant consideration of what is in the best interest of the students while also generating the best strategy to enlist buy-in from all teachers. I am a firm believer in the concepts of change that are outlined in CBAM (concerns-based adoption model) discussed by Gene Hall and Shirley Hord (2011). The authors reviewed several strategies to help leaders inventory the types of teachers under their domain and determine how each individual interprets change and, more importantly, how each individual responds to being asked to change. In my time as codepartment chair, I used Hall and Hord's recommendations to guide how I confronted teachers when informing them that I was considering starting a discussion about a change to a department policy or curricular element. The strategies I employed ranged from preelecting an individual privately before a department meeting so they could warm to the idea to being very direct in asking someone to support a new initiative. I learned that the critical piece to making change creditable was creating a space, physically and mentally, in which everyone put the students and mission ahead of personal interpretations or perspectives. This applies to everything from a change in office locations to implementing a piece of the *NGSS*. Change is only credible if it appears to be—and is—fully informed by the many forces that can potentially act upon the department during and after the implementation.

The science department is not unique in its need for collaboration and receptivity to change; however, the various beliefs that emerge from scientific knowledge and cultural interpretations create an atmosphere that requires leadership with a clear vision or mission. Furthermore, the leadership of a science department must be versed in strategies to engage all members in a way that focuses on student success, skill development, and an informed perspective.

Where Am I Today? Questions to Ask Yourself

For You as a Science Teacher

1. How would I describe my own science education at both school and at university?
2. Which curriculum tradition do I most identify with, and why do I hold these beliefs? How do my beliefs shape my teaching?
3. What do I believe about the nature of science, and how do these beliefs shape my teaching?
4. What do I think are the most important qualities to develop in all my students?
5. Have I ever *seriously* considered my beliefs about science and science education? What prompted these considerations?
6. What do I consider to be foundational pedagogical strengths of my teaching and in what areas do I know, intuitively, that I must improve?
7. How do I feel when told that I should embrace a particular reform?
8. What would stop me from *seriously* considering my beliefs about science and science education?
9. How might subject associations (for example, NSTA) help me rethink my beliefs on science and science education?

For You as a Department Chair

1. What is the history of my department in the school? What power does it possess?
2. How does the department project its power? Examples may include preferential timetabling and greater access to resources.
3. What opinions do senior administrators have of the department?
4. What views do my colleagues have of science and science education, and how do they view the work of the department? community, organization, or some combination of both?
5. How would you use the term *collegial* when describing your department?
6. If I consider promoting changes in science teaching and learning within the department, where would I find support and/or opposition?
7. How well do I know the teachers in my department, and their beliefs about science and science education?
8. What does a professional learning conversation look like in your department?

2

CHANGING SCRIPTS

In this chapter we progress from the historical development of the science curriculum to the difficulties that teachers face with the apparently endless calls to reform and the reasons those reforms continue. The press for reform will continue, and as science teachers we need to understand the very real motivations for this continued force. The call for science to be taught as inquiry goes back to the 1960s. If the practicing arm of our profession doesn't acknowledge—or perhaps resists—what the research arm is proposing, then the reform is bound to come up again. A wizened (or cynical, take your pick) vice principal once told one of us that if we stayed in teaching long enough, we would see the same reforms come around again, just relabeled. Twenty-five years later he has been proven correct.

As we said in Chapter 1, there is no point in attempting to blame teachers for not implementing the lofty ideals of the various reform documents. Change is hard, not just in teaching, but in all aspects of our lives. To even attempt to change the way that we teach, a way that we are heavily invested in as science teachers, requires that we first understand why we teach the way we do and then to believe that there are solid reasons for changing. From there, we must explore what the alternatives are to the way that we work and why they are realistic, and then be given the time and space to experiment with the alternatives to give them meaning and significance. An important point in all this is to recognize that, for an individual teacher to change within a traditional department is close to impossible. If we are to move beyond the cynical (but apparently accurate) observations of the vice principal mentioned above, then we need to reimagine departments as places where it is safe to challenge the academic tradition of science teaching and to experiment with alternatives that will help us to reform science education for the benefit of our students. So, to begin, let us consider the academic tradition and the power that it has over us as teachers. To do this, we turn to the concept of the teaching script.

The Power of the Script

A teaching script can be defined as a framework that teachers use to conceptualize the act of teaching. The script takes into account everything that happens within the classroom, from the role of the teacher to the role of the student; from the way that teachers and students interact, to what teaching and learning look like; from what teachers understand to be "good" science teaching to how students perceive science. Some have described the script "as the familiar narratives that help actors [teachers and students] involved in an interaction understand what they might expect through their engagement in the interaction" (van Langenhove and Harré 1999, pp. 351–352). The idea of a script might be made clearer by the sample script offered by Bain (2004), who describes what happens as students enter a large lecture hall for the first session of a college class. Typically, because students have had prior experiences in classrooms, they enter before the class begins, find a seat, and may get out a notebook to take notes. In this familiar script, all the seats face a specific direction—toward the point at the front of the room to which the professor will inevitably locate. Once located, the learned professor will use a chalkboard, a whiteboard, or a likely overly busy PowerPoint to dispense information. The cues in the script, from the position of the professor to what students expect from the scene, set the roles of both professor and students. Through the powers of socialization, the professor has expectations of the students that coincide with this script. It is expected that when the professor arrives, the students will already be in their seats and ready to listen and write as the instruction is provided. However, there might be subscripts that are cued as more information about the professor, the environment, and students' expected roles becomes clearer throughout the semester. As an example, a particular student might decide, "Oh, this is a course like the one I had with Professor X, where he expected students to solve problems in groups for a portion of the lecture." In this case, Professor X's class may become the subscript by which future lectures may be judged. If this subscript is aligned with the script in the new course, then the student will feel engaged. If the subscript does not align, then there is potential for disengagement. We have all experienced this: A colleague produces an innovative unit of work, and then students are dissatisfied when they are not similarly engaged by our teaching. Many more aspects of scripts could be highlighted, but we offer these examples to provide a clearer image of what we mean by a script and its role in social activities such as teaching and learning.

While scripts guide all aspects of our lives, it is likely that we use them without explicitly being aware (or critical) of them, even though they are central in helping us decide what to expect and how to act in different situations. It is important to note that a script does not predict exactly what will happen, but instead provides a loose framework for what is likely and feasible given past experiences. In this chapter specifically, and in the book more broadly, we share our belief that intentionally bringing the notion of script to the surface is important so that we, as science teachers, can become more cognizant and critical of such scripts. With this understanding, we move on to the predominant script of our learning and teaching of science, and explore why it is so difficult to change. After that, we focus on reforms that challenge this academic script, including the scripts proposed by the most recent national standards documents such as *A Framework for K–12 Science Education: Practices, Crosscutting Concepts, and Core Ideas* (NRC 2012) and the *Next Generation Science Standards* (*NGSS*; NGSS Lead States 2013). Finally, we will discuss why the department is the place where teachers can work together to achieve changes that will improve both the teaching and learning of science that happens in the school. It is not enough to hope for change, for that lacks intentionality. Improvements only happen if the department is reimagined as being the place where reform can occur.

The Academic Script and the Teacher

As we discussed in Chapter 1, the dominant script of science teaching we see in schools today has a long history and is firmly rooted in the values of the academic tradition:

> *Conceptual knowledge, compartmentalised into distinct disciplinary strands, the use of key, abstract concepts to interpret and explain relatively standard problems, the treatment of context as mainly subsidiary to concepts, and the use of practical work to illustrate principles and practices. (Tytler 2007, p. 3)*

It is important to acknowledge the academic script for three reasons: First, as teachers we have been successful in this form of science education. Second, having been successful as students, we tend to reproduce the teaching that worked for us. The third reason is that, while the academic script was successful for us, the relationship between science and society has changed.

We Were Successful

The first reason for understanding the academic script is intensely personal: As science teachers, *we have been successful in this tradition.* As high school students we chose science courses, attended class, carried out experiments, learned how to produce a lab report, and passed the tests and exams. For reasons of interest, enjoyment, economics, peer or family pressure, or for lack of another career path, we continued our science studies at a postsecondary institution. Lessons became lectures, and assessments consisted of a few major papers and lab reports. Most of the final evaluation rested on lengthy, comprehensive, rigorous exams. Through all of this we persevered, graduated, and decided to pursue a teaching degree. For some of us, a special teacher inspired us by making such a significant difference to our lives and learning that we decided to pursue a teaching career. Some of us may have thought we had a gift for teaching, while others may have been attracted by benefits such as salary and the naive perception of long summer holidays. Some of us became teachers because we didn't know what else to do with our science degree, or because market conditions did not allow us to enter our preferred occupation, and still others came to teaching later in life after beginning careers in other vocations and then finding them wanting for many different reasons.

Regardless, by the end of our university education, we all knew a little about a plethora of science-related disciplines and a great deal more specific knowledge from fields such as biology, chemistry, physics, environmental science, and so on. We were "experts" and certified by teacher certification boards to pass that expert knowledge on to the generations of high school students who would pass through our classrooms. The question was, how would we teach those students? This brings us to the second important reason for understanding the academic script. Being successful as a student in the academic script socializes us into it and thus shapes us as a teacher. In short, we will teach as we were taught because that is how we were successful.

Teaching as We Were Taught

We would argue that most science teachers who have entered the profession over the past 100 years have started their teaching careers in a manner that largely emulated the teaching and learning that they were socialized into as students. As teachers, we maintain the status quo—and by doing so increase the inertia resisting change. In our travels from elementary, to secondary, to postsecondary science it is clear that we "got it." We were successful, having learned how to play the science game, and the confirmation of our success was seen in our science and education degrees. These confirmed our membership in a body rich in academically focused instructional strategies. Those strategies (and our adherence to them) are so resistant to change that persistent calls for reform have "left relatively undisturbed the major narrative of the science curriculum that focuses on the establishment of a body of knowledge that is assessed largely by declarative means" (Tytler 2007, p. 3).

The strategies that support this major narrative remain as pervasive in classrooms today as they have since the days of the 1867 *Scientific Education in Schools* report. They are rooted in "memetic" activity (Jackson 1986), in which conceptions of learning require students to repeat (or mime) information back to the teacher in classroom discourse, reports, quizzes, or tests. Consequently, the following characteristics of science classrooms are familiar to us as both students and as teachers:

- The curriculum emphasizes basic skills.
- Instruction relies heavily on textbooks and workbooks.
- Student acquisition of information is prioritized.
- Students are seen as blank slates and information is etched onto these slates by the teacher.
- Scientific knowledge is presented in a didactic manner through lecture, text, and demonstration.
- Students are asked to recite acquired knowledge.
- Students are tested for factual information at the end of units or chapters (NRC 1996).

Working from this script, the students' role is to be filled with expert knowledge, while the teacher's role is commonly seen as being the authority for validating student understanding. Yager (2005) captures the heart of the danger of teaching as we were taught in these words:

> *Too often teachers as well as the general public are too willing to ignore the essence of science and to relegate its teaching to the topics too often characterizing curriculum frameworks, textbook chapters, the "agreed upon" concepts too often packaged in discrete disciplines. (p. 20)*

Teaching as we were taught is the comfort zone for many science teachers, but it ignores both the essence of science and the fact that the relationship between science and society has changed. This is the third reason that we have for needing to understand the academic script: Although we were successful, the academic script is increasingly alienating for our students. There is a flight from science.

A Changing Relationship

To be a science teacher is to be in a privileged position within a school. Science is seen as an important subject; it would be highly unusual for a science teacher to be asked, "How will this subject help my child get a job?" This importance endures despite the political pressure for standardized testing in numeracy and literacy in most English-speaking countries over the past two decades. The academic status that we have inherited is courtesy of the 19th-century push to professionalize science. As science teachers, we have our own language and research practices and a comprehensive knowledge base, and we recognize the dependence society has on the technological advancements such knowledge provides. Having been successful with the academic script, and strongly socialized into the power of science, we know what good science teaching looks like. Having been successful, we now want the same for our students. That is an honorable goal, but for us to continue to teach as we were taught would be a disservice to many of our students. If we teach as we were taught, then the only students we would reach would be those like us: the ones who get it, the ones who have been visible to you because you were one of them. That's a very important but very small number of students. Society needs an ever-increasing number of successful science graduates: We do not need to make the case here for creating more health, environmental, engineering, energy, and other such professionals. The problem, for the most part, lies in the way we teach, a way that while "designed principally to train young people as a preparation for entering the science discipline, is the very instrument that is turning them away from science" (Tytler 2007, p. 18).

We are sending fewer students in to science- and technology-based vocations and proportionally fewer secondary students are selecting the science subjects that will get them into science-related postsecondary programs. What's more, potential employers are finding that those who do graduate are short of competencies such as critical-thinking skills, collaborative work skills, the ability to apply concept knowledge creatively in novel situations, and problem-solving ability. In 2013, the psychologist Dean Keith Simonton argued in the journal *Nature* that science will rarely see individual acts of genius in the future; rather, it will rely on the capacity to work in collaborative teams (Simonton 2013). The competencies that we have listed here are also seen to be integral to success in both domestic and global competitive markets, a catalyst for so many of the educational reforms in the areas of science, mathematics, and engineering.

Science may be respected by mainstream society, yet it is increasingly seen as an inappropriate career choice by students. Too many of our students drop out of science by the end of grade 10 and in so doing eliminate science-related college and apprenticeship programs as potential future employment opportunities. There are multiple reasons for this increasing disconnect between science education in schools and science, and these are summarized by Tytler (2007, pp. 3–4) as the changing practice of science, the changing nature of public engagement with science, public challenges to science, the knowledge explosion, and changes to the nature of schooling.

The first consideration is the changing conduct of science: Gone are the days of the lone genius working on his or her own or in close correspondence with other scientists. Science is increasingly globalized, multidisciplinary, technology heavy, funded by governments and private industry (with implications for control and dissemination), and scrutinized and challenged by other agendas. All of these changes have implications for the representation of science and the skills and attitudes that our students require if they are to pursue science-based careers.

Second, public engagement with science has never been higher, if the reporting and analysis of science-related issues in the media is any guide. The public is rightly concerned with the applications of science and technology, the impacts of science on society and the environment, and the decision-making processes that accompany many issues. Science education has a key role in equipping students with the capacity to make informed decisions about any number of science-related issues that affect them and society.

Third, and closely related to public engagement with science, are the increasingly strident challenges to science. These range from critiques of the power and status of science, involving who determines scientific knowledge and who may access it, to questions of the nature of science in relation to indigenous ways of knowing, values, society, and the environment. Closely allied to these challenges is a growing sense of distrust of science perpetuated by those with other agendas, such as the continuing claims of controversy within the scientific community over climate change. Our students need to be equipped to handle such issues, while maintaining a healthy skepticism toward scientific claims.

Fourth, scientific knowledge, and access to it, has grown exponentially since 1945. The science curriculum in schools, therefore, is always playing catch-up with evolving concepts, issues, and discoveries. The implication is that for many people decisions are being made, at both the personal and societal levels, in the absence of adequate scientific knowledge. Examples include reproductive technologies, resource utilization, and the appropriate application of technology. In addition to this, access to scientific knowledge is rapidly evolving, with the capacity to search online for information (which may or may not be peer reviewed) potentially calling into question the knowledge expertise of science teachers.

The "teacher as expert" model is predicated on delivering a stable body of knowledge to students who would otherwise have difficulty accessing that knowledge. Disciplinary knowledge is important, but if students can access that information outside the classroom, then the role of the teacher is fundamentally challenged. In response, the focus for science teaching needs to be on facilitating learning and developing the skills, attitudes, and abilities that support student learning. This is particularly important in terms of the changes to the student population over the past few decades.

As we have seen, science education has been dominated by the academic tradition since the 1850s. The huge expansion in access to secondary schooling since that time means that we have a curriculum unsuited to those who are not well served by the academic tradition. Here is an unresolved tension at the heart of contemporary science education: How do we meet the need to educate the next generation of science professionals, while at the same time providing all students with a comprehensive science education that allows them to make informed personal and societal decisions when engaging with issues that require a knowledge of science? Given the changing nature of young people, this question becomes even more acute.

The values that the majority of contemporary teachers hold to are not the values that young people in postindustrial societies are articulating to researchers. Whereas many of us would hold to a more traditional view of personal economic stability and career development, young people are increasingly concerned with issues of uncertainty, change, sustainability, and community. This translates into a preference for flexibility over stability and an individual concern for making proactive decisions that promote personal autonomy and the capacity to respond to changing conditions. Such a value set requires a different set of skills, attitudes, and abilities to those promoted by a more traditional conceptualization of the world.

In summary, the problem with the academic script lies in the fact that it has not evolved in response to these changes—and part of that phenomenon lies in the fact that we have been successful with the academic script as it stands now. To challenge the academic script is difficult, there is no doubt about that, but do not despair! Teachers are not entirely powerless in being a force for change: "Whatever curriculum writers, politicians, administrators, and academics say are the aims of the course, what actually happens is in the hands of teachers" (White 1988, p. 10).

In the next section we consider reform efforts that have developed in response to the dominant academic script, starting with the work of Joseph Schwab and leading up to the most recent efforts. Documents such as the *NGSS* hold great promise for change, but to be effective, teachers will need to do more than read them: Teachers will need to work with them and come to understand what they mean in the context of their own classes and schools. And the work of giving any reform a sense of meaning has to occur in the department, for only the department has the potential or capacity to be a collaborative community of learners.

Reforming the Script

Even as modern science education began to evolve, there were disparate views as to what purposes it should serve. Challenges to what became the academic script have existed from the days of Reverend Dawes. These challenges all share a common denominator: a commitment to giving students experience *with* science and utilizing that experience to develop, explain, and communicate explanations that are supported by scientific evidence. The implication is that teachers who have been successful with the academic script will need to reconsider their relationship to that script and work to give meaning to the reformed scripts.

In this section, we will consider the major reform efforts of the past half century and how they have evolved, for in understanding that evolution, we can appreciate that the underlying tenets of the reforms remain constant. The wheel is not being reinvented, rather it is being refined—as many theories are over time; why should science pedagogy be any different? As teachers and chairs, this should give us confidence that efforts to reimagine our departments will be supported into the long-term future.

From Schwab to the National Science Education Standards

Joseph Schwab, in his widely influential piece titled "The Teaching of Science as Enquiry" (1958, 1962), described the origin of the academic script as being rooted in a naively literal interpretation of science. In Schwab's view, the processes of science were directed toward observing natural phenomena and then recording them as inalterable truths. From this interpretation of science, Schwab (1958) outlined a framework for education in the sciences:

> *The education appropriate to such a view of science was clear enough: mastery of the true facts as known by science. For such an education, the best possible material was one kind only: a clear, unequivocal, coherent organization and presentation of the known: a pure rhetoric of conclusions. For neither doubt nor ambiguity characterized what was known. A declarative rhetoric of conclusions, omitting all evidence, interpretation, doubt, and debate sufficed. For, presumably, no interpretation was involved, no doubt existed. The conclusions of science merely presented what the scientist had seen.* (p. 375)

Schwab saw the academic script described in this way as an authoritative script. The academic script focused on lectures, textbooks, and demonstrations in which the conclusions of science were prioritized but not problematized. Laboratory experiences were part of instruction, dictating what students were to look for and exemplifying the precise conclusions that supported the canonical truths of lectures and textbooks. Figure 2.1 (p. 30) reveals a worksheet section from a study guide, common in many contemporary classrooms, that can be interpreted as very much aligned with what Schwab described as instruction in the inalterable truths of science.

In the Figure 2.1 worksheet, declarative knowledge is emphasized without reference to evidence, interpretation, doubt, or debate. Furthermore, whether intentional or not, this worksheet is founded on declarative knowledge or conclusions from science and absent of the processes of science that contextualize it, hence conveying a powerful message to students about the enterprise of science.

Figure 2.1

WORKSHEET USED TO GUIDE INSTRUCTION AND TEST PREPARATION

Populations worksheet

Genes: A structure in an organism's cells that carries its hereditary information
Extinction: The disappearance of all members of a species from earth
Endangered: A species that is in danger of becoming extinct
Threatened: A species that could become endangered in the near future

What is the major cause of extinction? Habitat destruction
Poaching: Illegal hunting of wildlife
Captive breeding: The mating of endangered animals in zoos for preservation
A plant's ability to fight disease is a result of its: adaptations to the environment
Why are rain forest plants sources of medicine? Some chemicals that rain forest plants produce to protect their leaves and bark can also fight human disease

While Schwab (1958) was explicit that the enterprise of science is distinctly different and more complex than the academic tradition would lead us to believe, he also realized that this has implications for the script that guides our teaching. Schwab pointed out, among other things, that declarative scientific knowledge, in addition to being bound by special reference, has a revisionary and plural character. He explained that the conclusions of science were really instances of phenomena that were special cases that hold true (or are consistent) with their characterization under certain conditions. As an example, he described the characterization of a newly isolated highly active element or isotope. In this example, understanding the new element or isotope involved understanding the parameters and methods of inquiry that led to the isolation in the first place. Schwab (1958) went as far as to declare that,

> *in general, the conclusions of scientific investigations are not about some quantity of nature taken in its pristine state, but about something which the principles of investigation have made, altered, or confined. The conclusions, therefore, are unintelligible or misleading unless they are known in the context of inquiry which structured and bounded the matters to which they refer. (p. 375)*

In describing the revisionary nature of scientific knowledge, Schwab sought to explain how scientific investigations necessarily required bounding the subject of inquiry, or studying a narrow instance of the subject, because a subject without bounds possesses much more richness and variety than can be reasonably understood. Using historical inquiries into the nature of light, Schwab chronicled how in 1665 Hooke described a wave theory of light that required thorough revision after Van Laue investigated X-ray scattering in 1923. Here was a differently bounded investigation of light, an example of the revisionary nature of the scientific enterprise. In this example, the revised scientific knowledge about light was not seen as more "right" but as "more comprehensive, more discriminating, or more nearly exhaustive of the originating subject materials" (Schwab 1958, p. 375). In this sense, a more realistic characterization of the enterprise of science reveals the tentative nature of scientific knowledge, in which knowledge claims are revisited and revised as new investigations explore additional instances, hence requiring more generalizable claims that are, in most cases, revisionary in nature.

Finally, Schwab described the pluralistic nature of the enterprise of science. The richness of the natural world, he argued, makes it possible that a specific phenomenon may represent multiple scientific principles, such that each principle cuts across the phenomenon in a distinct and unique way. In more recent discussions of the nature or enterprise of science, this is more fully understood as science being "theory laden." Lederman (2007) explains this:

> *Observations (and investigations) are motivated and guided by, and acquire meaning in reference to, questions or problems. These questions or problems, in turn, are derived from within certain theoretical perspectives. Often, hypothesis or model testing serves as a guide to scientific investigations. (p. 834)*

The new professionalizing science of the 19th century was built on a solid foundation of observation, recording, and categorizing the physical world. Mathematics, the queen of the sciences, became increasingly prominent as time progressed, leading to a more complex notion of the enterprise of science, reliant on frameworks, in the form of theoretical perspectives, as well as questions or problems to help guide what was observed.

Given this more complex notion of the enterprise of science (i.e., notions about the special reference, the revisionary and plural nature), Schwab (1958) proposed that the academic script needed to be revisited and reformed to better reflect science. Instead of a pure rhetoric of conclusions conveyed to students by teachers through authoritative lectures and determinations, and in the absence of attention to evidence, interpretation, doubt, and debate, he proposed a rhetoric of investigation. Through the rhetoric of investigation, science is presented as inquiry in which students engage in investigations to learn. Schwab realized that this also meant a change in what was to occur in classrooms and laboratories. As an example, instead of a laboratory focused on validating the "known truths" presented in class, students would use laboratory time to conduct mini-inquiries. Instead of focusing on finding the right answers, students would spend time formulating problems and methods for gathering evidence that could be used to formulate reasoned explanations to solve these problems. Schwab (1958) described how the students' mini-inquires could be situated among historical inquiries by engaging in early investigations of historical import (e.g., the development of the Bohr model of the atom). The assumptions and principles of the time could be introduced within the context of student inquires, allowing the construction of explanations that coordinate the evidences and theories of the historical point. These explanations are then open to challenge and debate in light of the explanations of the time period. Schwab explained how these proposed inquiries enabled students to learn the bounded principles demonstrated by historically isolated phenomena and subsequently apply this knowledge to their own inquiries. Importantly, Schwab revealed how these proposed changes to the academic script positioned students to develop a more realistic conception of the enterprise of science that incorporated the referential, revisionary, and plural nature of science.

What emerged from the 1960s, grounded by Schwab's work, was a call for a broadened emphasis in science teaching so that science processes, which had previously been confined to learning about the linear, sequential, and one-directional scientific method, were prioritized equally alongside the learning of key concepts or science principles. For these reformers, the scientific method was more than simply a component of the authoritative rhetoric of conclusions; students were to actively engage with basic scientific processes such as observing, clarifying, measuring, inferring, and predicting (Bybee 2011). Additionally, instructional materials were created during the 1960s and 1970s that sought to support students' laboratory experiences. The aim of these materials was to reunite the processes and products of the scientific enterprise, thus developing deeper understandings of key concepts and science principles. Teaching "science as inquiry" was born.

Reformers' support for teaching science as inquiry continued to grow from the 1960s to the 1990s, as evidenced by the national science standards documents of the 1990s and 2000s. In addition, position statements from leading national science education organizations promoted inquiry as a central instructional strategy for teaching science (AAAS 1993; NRC 1996, 2000, 2005; NSTA 2007). These calls for reform are exemplified in the declaration by the National Science Teachers Association (NSTA) in 2007 that "inquiry-based laboratory investigations at every level should be at the core of the science program and should be woven into every lesson and concept strand."

The high-powered promotion of the reforms might imply that teaching science as inquiry became commonplace in science classrooms in the half century since the work of Schwab. As we know, this is not the case; most evidence suggests that it did not (NRC 2005; O'Sullivan and Weiss 1999). The limited implementation has been due to confusion as to how it could be operationalized in classrooms, what was actually meant by inquiry, concerns over materials, time and curriculum coverage, and most tellingly, by a concern that the traditional emphasis on canonical knowledge was being diluted. As we saw in Chapter 1, calls for reform tend to ignore the resilience and power of the academic script. That script has been developed over close to 200 years by powerful historical, political, social, economic, and educative forces. Teachers have been successful with this script, in the process becoming socialized into it, and replicated it in their own teaching. Consequently, teachers find it difficult to reshape their instruction, a situation that has not been helped by the lack of effective professional development. On that somewhat discouraging note, we now turn our attention to the next iteration of reforms to science teaching, the *NGSS*.

Next Generation Science Standards

A word of caution before we start. Our intention here is twofold: to place the *NGSS* (and the complementary *Framework* document) in the historical context of reforms to science education and to give a brief outline of the major features of these documents. While this information is useful, it is not a substitute for actively engaging with the document and, over time, developing your expertise. What we offer here is a guide to getting started—but you need to take the first (and sometimes tentative) steps. If you are new to these documents, a good place to start on the journey is *www.nap.edu/catalog.php?record_id=13165* for the *Framework* and *www.nextgenscience.org* for the *NGSS*.

Over the last decade, Schwab's ideas around the teaching of science as inquiry have been expanded and enriched by a renewed focus on the practices of science. The result of this work is to be found in the most recent standards documents: the *Framework* and the *NGSS*, which are founded on earlier works such as *Taking Science to School* (Duschl, Schweingruber, and Shouse 2007) and *Ready, Set, Science!* (Michaels, Shouse, and Schweingruber 2008). These two books are themselves built on standards documents of the 1990s (i.e., AAAS 1993; NRC 1996, 2000) and further informed by science teaching and learning research since then. *Taking Science to School* and *Ready, Set, Science!* provided an enriched and expanded framework for science teaching and learning that set the stage for what followed in the *Framework* and *NGSS* documents. In particular, *Ready, Set, Science!* explained that the basis of this enhanced framework

> *rests on a view of science as both a body of knowledge and an evidence-based, model-building enterprise that continually extends, refines, and revises knowledge. This framework moves beyond a focus on the dichotomy between content or knowledge and process skills, recognizing*

> *instead that, in science, content and process are inextricably linked.* (Michaels, Shouse, and Schweingruber 2008, p. 17)

Taking Science To School and *Ready, Set, Science!* both describe four strands of science that inextricably link science content and process. Students should be able to

- know, use, and interpret scientific explanations of the natural world;
- generate and evaluate scientific evidence and explanations;
- understand the nature and development of scientific knowledge; and
- participate productively in scientific practices and discourse.

Finally, these documents also introduced a focus on practices of science instead of inquiry, with Michaels, Shouse, and Schweingruber (2008) proposing the following explanation:

> *We talk about "scientific practices" and refer to the kind of teaching that integrates the four strands as "science as practice." Why not use the term "inquiry" instead? Science as practice involves doing something and learning something in such a way that the doing and learning cannot really be separated. Thus, "practice" ... encompasses several of the different dictionary definitions of the term. It refers to doing something repeatedly to become proficient (as in practicing the trumpet). It refers to learning something so thoroughly that it becomes second nature (as in practicing thrift). And it refers to using one's knowledge to meet an objective (as in practicing law or practicing teaching).* (p. 34)

Working on the foundation provided by Schwab, the standards and later publications have all contributed to shape the framework for the *NGSS*, which recommend that science teaching and learning should rest on the following three dimensions:

- Scientific and engineering practices
- Crosscutting concepts that unify the study of science and engineering through their common application across fields
- Core ideas in four disciplinary areas: physical sciences; life sciences; Earth and space sciences; and engineering, technology, and applications of science

More specifically, this has been referred to as three-dimensional learning, whereby learning is engaging in science and engineering practices to use disciplinary core ideas and crosscutting concepts to explain phenomena or solve problems.

While there are many unique features to these dimensions, we wish to foreground the science practices described in the first dimension, only because we see instruction grounded in practices to explain phenomena or solve problems as most challenging to the traditional script of teaching science. The emphasis on practices reflects how Schwab's notion of science as inquiry has been developed in response to the changing relationship between science and contemporary society. Given the resiliency of the academic script, a focus on the practices of science highlights the difficult, though important, work ahead for teachers as they reexamine their own teaching practices, and the role that the department can play in supporting that work. So to start, the question that must be asked is, what do we mean when we talk of the practices of science? For without asking questions, we can never start to critique our work and see how it needs to change to promote the teaching and learning of science in our classrooms.

Practices can be described as "a set of sensible actions that are both performed by members of a community and that evolve over time" (Berland 2011, p. 627). The *Framework* offers the following eight practices, which are consistent with Berland's (2011) description of a set of actions performed by the scientific community that are evolving over time:

- Asking questions (for science) and defining problems (for engineering)
- Developing and using models
- Planning and carrying out investigations
- Analyzing and interpreting data
- Using mathematics and computational thinking
- Constructing explanations (for science) and designing solutions (for engineering)
- Engaging in argument from evidence
- Obtaining, evaluating, and communicating information

Additional information about each of these eight practices is described in further detail in Table 2.1 (p. 36).

So, consistent with what Schwab (1958) proposed as the benefits of teaching science as inquiry, Bybee (2011) offers a rationale for the recent focus on scientific practices:

> *When students engage in scientific practices, activities become the basis for learning about experiments, data and evidence, social discourse, models and tools, and mathematics and for developing the ability to evaluate knowledge claims, conduct empirical investigations, and develop explanations. (p. 38)*

Table 2.1

SCIENTIFIC AND ENGINEERING PRACTICES OUTLINED BY NRC

PRACTICE	DESCRIPTION
ASKING QUESTIONS (FOR SCIENCE) AND DEFINING PROBLEMS (FOR ENGINEERING)	Science begins with a question about a phenomenon, such as "Why is the sky blue?" or "What causes cancer?," and seeks to develop theories that can provide explanatory answers to such questions. A basic practice of the scientist is formulating empirically answerable questions about phenomena, establishing what is already known, and determining what questions have yet to be satisfactorily answered. Engineering begins with a problem, need, or desire that suggests an engineering problem that needs to be solved. A societal problem such as reducing the nation's dependence on fossil fuels may engender a variety of engineering problems, such as designing more efficient transportation systems, or alternative power generation devices such as improved solar cells. Engineers ask questions to define the engineering problem, determine criteria for a successful solution, and identify constraints.
DEVELOPING AND USING MODELS	Science often involves the construction and use of a wide variety of models and simulations to help develop explanations about natural phenomena. Models make it possible to go beyond observables and imagine a world not yet seen. Models enable predictions of the form "if … then … therefore" to be made to test hypothetical explanations. Engineering makes use of models and simulations to analyze existing systems so as to see where flaws might occur or to test possible solutions to a new problem. Engineers also call on models of various sorts to test proposed systems and to recognize the strengths and limitations of their designs.
PLANNING AND CARRYING OUT INVESTIGATIONS	Scientific investigation may be conducted in the field or the laboratory. A major practice of scientists is planning and carrying out a systematic investigation, which requires the identification of what is to be recorded and, if applicable, what are to be treated as the dependent and independent variables (control of variables). Observations and data collected from such work are used to test existing theories and explanations or to revise and develop new ones. Engineers use investigation both to gain data essential for specifying design criteria or parameters and to test their designs. Like scientists, engineers must identify relevant variables, decide how they will be measured, and collect data for analysis. Their investigations help them to identify how effective, efficient, and durable their designs may be under a range of conditions.

Table 2.1 (*continued*)

PRACTICE	DESCRIPTION
ANALYZING AND INTERPRETING DATA	Scientific investigations produce data that must be analyzed to derive meaning. Because data usually do not speak for themselves, scientists use a range of tools—including tabulation, graphical interpretation, visualization, and statistical analysis—to identify the significant features and patterns in the data. Sources of error are identified and the degree of certainty calculated. Modern technology makes the collection of large data sets much easier, thus providing many secondary sources for analysis. Engineers analyze data collected in the tests of their designs and investigations; this allows them to compare different solutions and determine how well each one meets specific design criteria—that is, which design best solves the problem within the given constraints. Like scientists, engineers require a range of tools to identify the major patterns and interpret the results.
USING MATHEMATICS AND COMPUTATIONAL THINKING	In science, mathematics and computation are fundamental tools for representing physical variables and their relationships. They are used for a range of tasks, such as constructing simulations, statistically analyzing data, and recognizing, expressing, and applying quantitative relationships. Mathematical and computational approaches enable predictions of the behavior of physical systems, along with the testing of such predictions. Moreover, statistical techniques are invaluable for assessing the significance of patterns or correlations. In engineering, mathematical and computational representations of established relationships and principles are an integral part of design. For example, structural engineers create mathematically based analyses of designs to calculate whether they can stand up to the expected stresses of use and if they can be completed within acceptable budgets. Moreover, simulations of designs provide an effective test bed for the development of designs and their improvement.
CONSTRUCTING EXPLANATIONS (FOR SCIENCE) AND DESIGNING SOLUTIONS (FOR ENGINEERING)	The goal of science is the construction of theories that can provide explanatory accounts of features of the world. A theory becomes accepted when it has been shown to be superior to other explanations in the breadth of phenomena it accounts for and in its explanatory coherence and parsimony. Scientific explanations are explicit applications of theory to a specific situation or phenomenon, perhaps with the intermediary of a theory-based model for the system under study. The goal for students is to construct logically coherent explanations of phenomena that incorporate their current understanding of science, or a model that represents it, and are consistent with the available evidence. Engineering design, a systematic process for solving engineering problems, is based on scientific knowledge and models of the material world. Each proposed solution results from a process of balancing competing criteria of desired functions, technological feasibility, cost, safety, aesthetics, and compliance with legal requirements. There is usually no single best solution but rather a range of solutions. Which one is the optimal choice depends on the criteria used for making evaluations.

Table 2.1 (*continued*)

PRACTICE	DESCRIPTION
ENGAGING IN ARGUMENT FROM EVIDENCE	In science, reasoning and argument are essential for identifying the strengths and weaknesses of a line of reasoning and for finding the best explanation for a natural phenomenon. Scientists must defend their explanations, formulate evidence based on a solid foundation of data, examine their own understanding in light of the evidence and comments offered by others, and collaborate with peers in searching for the best explanation for the phenomenon being investigated. In engineering, reasoning and argument are essential for finding the best possible solution to a problem. Engineers collaborate with their peers throughout the design process, with a critical stage being the selection of the most promising solution among a field of competing ideas. Engineers use systematic methods to compare alternatives, formulate evidence based on test data, make arguments from evidence to defend their conclusions, evaluate critically the ideas of others, and revise their designs to achieve the best solution to the problem at hand.
OBTAINING, EVALUATING, AND COMMUNICATING INFORMATION	Science cannot advance if scientists are unable to communicate their findings clearly and persuasively, or learn about the findings of others. A major practice of science is thus the communication of ideas and the results of inquiry—orally, in writing, with the use of tables, diagrams, graphs, and equations, and by engaging in extended discussions with scientific peers. Science requires the ability to derive meaning from scientific texts (such as papers, the Internet, symposia, and lectures), to evaluate the scientific validity of the information thus acquired, and to integrate that information. Engineers cannot produce new or improved technologies if the advantages of their designs are not communicated clearly and persuasively. Engineers need to be able to express their ideas, orally and in writing, with the use of tables, graphs, drawings, or models and by engaging in extended discussions with peers. Moreover, as with scientists, they need to be able to derive meaning from colleagues' texts, evaluate the information, and apply it usefully. In engineering and science alike, new technologies are now routinely available that extend the possibilities for collaboration and communication.

Source: NRC 2012, pp. 50–53

Important for us, unlike when Schwab (1958) proposed teaching science as inquiry, the focus on engaging students in scientific practices comes at a time when research is well-situated to inform teachers as to how to begin challenging the academic script and thus begin to develop a new script. This is significant, for we as teachers need more than words on a page to guide changes in our teaching and learning. To question is an excellent starting point, but for real change to occur and be sustained, the questioning must spark long-term conversations about the meanings we apply to phrases or words such as "practices."

One example of the research that will help teachers refocus their instruction on the practices of science is the work on engaging students in developing and using scientific models. This work has begun to reveal and describe instructional approaches that can lead students to explore scientific phenomena, help students express their initial and developing conceptions about scientific phenomena using disciplinary core ideas and crosscutting concepts, guide experimentation to refine those models that serve as explanations, and engage students in argumentation about scientific phenomenon, as competing models are evaluated on the basis of their utility as explanations and consistency with the specific scientific phenomena (Campbell, Oh, and Neilson 2012). Additionally, NSTA—in its K–12 practitioner journals *Science and Children*, *Science Scope*, and *The Science Teacher*—has begun to highlight research about science and engineering practices as they translate into instructional practice in a move that offers support to teachers looking to align their teaching with the reforms (e.g., for a discussion of asking questions, see Bell et al. 2012; for developing and using models see Krajcik and Merritt 2012; and for using mathematics and computational thinking see Mayes and Koballa 2012). In summary, research into the teaching and learning of practices shows promise for supporting a reformed script that more accurately reflects the work of scientists and the enterprise of science. However, as with any change, we can can also identify factors that will inhibit these changes, most of which all of us have come across in our careers, sometimes repeatedly.

Limiting Learning?

There is a gradual shift away from a "one size fits all" approach to professional development toward opportunities that emphasize teachers' active learning of content and pedagogy that align with subject-specific standards. This is a positive development because teachers who are looking to question their current work face a number of challenges. These include limited conceptions of the nature of science and a consequent constriction in their teaching repertoires, a lack of content knowledge, inexperience with the range of reform-oriented teaching approaches, and an inadequate (given their own educational background) understanding of science and engineering practices. For example, when using scientific models in their classroom instruction, there is evidence suggesting that teachers, who themselves have had little experience in developing and using models, use them in ways that are inconsistent, framed by underdeveloped views, or simply very limited in their application (van Driel and Verloop 1999). Again, we must stress that this is not meant to disparage teachers or teacher preparation programs, but instead to highlight the complexity that each of these practices within the three-dimensional learning framework, as just one component of the *NGSS* teaching script, brings with it. Given the realization of this complexity, the NRC asserts that "teachers are the linchpin in any effort to change K–12 science education … [and] the professional development of teachers of science will need to change" (NRC 2012, p. 255). Professional learning, the main focus of this book, will be central to any change to the long-standing academic script. Therefore, we need to devote some time critiquing contemporary professional development practices and their impact on professional learning.

Let us be brutally honest here: Historically, a great majority of professional development has resulted in limited teacher professional learning. The main reason for this is the disconnect between the professional development delivered in workshops and the realities of conditions and expectations in classrooms. Darling-Hammond and Sykes (1999) comprehensively fault current practices as

> *focusing on district-mandated, generic instructional skills of teachers "trained" as individuals by an outside "expert" away from their job site. Because this training is fragmented, piecemeal, and often based on instructional fads, it is viewed as a frill, easily dispensed with in tough financial times. Perhaps most damaging, these workshops, although they often respond to expressed teacher needs, are seldom explicitly linked to what schools expect students know and be able to do.* (p. 134)

This disconnect can only be resolved if efforts are made to link or support what teachers are learning (i.e., new materials, methodologies, and practices, such as those found in the *NGSS*) with the everyday classroom experiences of teachers. This need not be more work, just a different form of work. Seamlessly linking professional development, in which teachers' experiences are central to learning, allows teachers to examine the basis of what is being learned in their classrooms and departments, the very places it will be employed. In this, teachers' prior knowledge about teaching as well as their experiences are seen as central to the development of new understandings and practices, such as those aligned with a reformed script for science education.

Finally, and this point serves as the connection between this chapter and the next, professional learning is not an individual activity. Professional learning is at its most effective when it is focused by teachers on their work and situated within the context of their workplace. The overwhelming majority of science teachers already work in departments, so professional learning opportunities can use a range of activities that already exist: coteaching, coplanning, or reflecting on lessons after they are enacted. The principal benefit of these types of interactions and subsequent professional learning is that it allows teachers to take a longer-term view of learning. The NRC (1996) emphasized the importance of teachers collaborating and taking responsibilities for their professional learning. This means teachers taking the initiative to become leaders, the instigators of change, and the producers of knowledge about teaching. It also means accepting the responsibility for deciding the content, the planning, and the conduct of activities, especially if we expect professional learning to meet the very real and specific needs of teachers in classrooms as they work to help enact and continuously refine a reformed teaching script. The role of the chair in reimagining the department as being the perfect place to support teacher professional learning is the focus of Chapter 3.

Summary

- To attempt to change the way we teach requires that we first understand why we teach the way we do and to believe that there are solid reasons for changing.
- It is important to acknowledge the academic script for teaching science and the impact it has on our own teaching.
- Teaching as we were taught often ignores both the essence of science and the fact that the relationship between science and society has changed.
- Documents such as the *NGSS* hold great promise for change, but to be effective teachers will need to do more than read them. Teachers will need to work with them and come to understand what they mean in the context of their own classes and schools.
- Since 1962, Schwab's ideas of science as inquiry have been expanded and enriched by a renewed focus on the practices of science. The result of this work is to be found in the most recent standards documents.

- Research in the teaching and learning of practices shows promise for supporting a changed script for teaching science as one that more accurately reflects the work of scientists and the enterprise of science.
- The professional development opportunities of science teachers need to change to support the implementation of the most recent reform documents.
- Professional learning is not an individual activity. Professional learning is at its most effective when it is focused by teachers on their work and situated within the context of their workplace.

Vignette 2

DAVID WELTY

David Welty has been teaching science at Fairhaven High School, Fairhaven, Massachusetts, for 14 years as a dual-certified biology and chemistry teacher. He has taught at all levels and currently teaches AP biology, AP chemistry, and college prep chemistry. In addition, he has served the science department for 12 years under many evolving titles from chair, to coordinator, and currently as teaching and learning supervisor. We have worked with David in the past and know that he is committed to cultivating a science department that fosters professional learning. On the strength of his work, we asked him to answer a few questions about his department that might offer some insights into how other chairs and departments navigate the topics raised in this chapter. These are the questions that David was asked, and having read his vignette, you might consider your own answers to them:

- *How would you describe the traditional script as it appears in your department, and is there a desire to change it? If not, why not? If yes, what has been the catalyst?*
- *What do terms such as* inquiry *and* science practices *mean in your department, and do these terms need to be clarified in practice before they are understood?*
- *What opportunities are there for professional development, and how effective have they been?*
- *How does (or how could) your department take the initiative in professional development?*

We often talk in our science department about what we think makes us successful teachers. It usually comes down to our own struggles in science: We struggled but we enjoyed the wonder of discovery, so we found ways to learn. We were successful because we worked to find ways to be successful. One consequence of this has been to transfer this approach to learning to our students, a strategy that has met with limited success.

I have thought students are not as interested in science as my peers were in the late 1980s, but upon reflection I count only 10 out of 200 former classmates who pursued the sciences. The same ratio holds in my high school's graduating classes today. Overall, we practice direct instruction because this is how we were successful. We try to make it active and more student centered, but the mindset of the students often appears to be that the teacher's job is to instruct, and if the students are not learning, then it is also the fault of the teacher. Teachers still often blame a lack of motivation and the work ethic of students for poor performances, but we need to do a better job of engaging with students and helping them recognize the wonder and opportunities that science presents. We need to move beyond how we were taught and develop strategies to move students from extrinsic to intrinsic motivation. The more successful teachers are also the better entertainers and can captivate the attention of the students. As a department, we are still searching and experimenting with lesson plan design and materials for better direct instruction. We believe that if we could construct more engaging minds-on and hands-on lessons, then more students would be intrinsically motivated and would embrace the wonder of science, perform better in science, and pursue science.

As a department, we are split on inquiry. Students are more engaged in classes where the focus is on inquiry, but the efficiency of transfer from learning activity to concept mastery seems limited. I believe in inquiry and try to embed it into my lessons either as a hook or a summary, but the meat of my instruction is direct instruction, active participation, checking for correctness, and practice. I remember my science and math teachers would put up a sample problem, demonstrate how to do it, and then put up another similar problem and have us do it. For example, my biology teacher would write a string of RNA codons on the blackboard and show how to translate the amino acids. I currently still use this approach.

We are now in our second year of being a professional learning community (PLC). As the teaching and learning supervisor, I am working to overcome the perceived discrepancy between using inquiry and gaining mastery. The science department is beginning to address the poor transfer of inquiry to mastery through conversation and formative assessment. Through our PLC, we have increased conversations about how to better deliver inquiry activities. We share what does or does not work and develop activities to improve our delivery. However, the biggest change has come with moving to more frequent formative assessments. The reality is that our 55-minute periods prevent entire laboratory activities to be completed—from prelab, to lab, to conclusion—in one day. A typical lab has become a three-day process. Consequently, we have started to use more formative assessments before, during, and after the activity to better structure our lab work and achieve the outcomes that are important to us.

In Massachusetts, the *NGSS* are being deferred until 2015, so we are (at the time of this writing) in a holding pattern. As a department, we explored the engineering standards of the *NGSS* in our PLC and brainstormed what, where, and when to implement. We are nervous about how engineering will fit into our core science lab courses, but overall see the *NGSS* as an opportunity to include more on the nature of science. We hope there will be specific professional development time to incorporate engineering into our core lab courses. Our main fear is not having the time to do a successful engineering–science inquiry unit with the current time demands to get through the state frameworks. If the state can free up time in the curriculum by reducing the content coverage when they adopt the *NGSS*, then the inclusion of the nature of science would be seen as more realistic. If time is not available, then coverage of material covered on state mandated exams will still be the highest priority. This would impact grades 5–10. For grades 11 and 12, space in the curriculum will have to be made to remove some content for more consideration of the nature of science. This would follow the improvements already implemented in the AP sciences.

Where Am I Today? Questions to Ask Yourself

For You as a Science Teacher

1. In what ways do I teach as I was taught?
2. Which facets of the academic script have I had the most success in challenging, and what has led to this success?
3. To what extent have I engaged in discussions about the newest standards documents with colleagues in my department? How might I start such conversations?
4. What facets of the newest standards documents have challenged me the most in terms of conceptualizing what is meant and how this will translate into experiences for students in my classroom?
5. What challenges do I expect might emerge when working to engage students in practices of science?
6. What would stop me from taking initial steps in engaging my students in science practices? What can I do to mitigate these obstacles?
7. What NSTA resources might help me move the ideas from the newest standards documents into practice in my classroom? (e.g., *The Science Teacher* articles, NSTA Learning Center).

For You as a Department Chair

1. Historically, what opportunities for professional learning in my department have been most successful, and what facets of these experiences can I help my colleagues consider for framing our work to shape a science practices script for science teaching and learning?
2. What are the current strengths of my department, and how can these strengths play a central role in professional learning focused on the newest standards documents?
3. What can the department do to ensure school administrators support commitments to science practices?
4. How can the work of the department be refocused in such a way that targets of the newest standards documents build on what we have already accomplished as a department?
5. What support is already in place for the newest standards documents inside and outside of the department?
6. What can I do to help make teachers' ideas about the newest reform documents more explicit so that they can be questioned and strengthened?
7. What structures and scaffolding do I have in place, or desire to develop, to have the kinds of professional conversations that provide teachers with a plan to create change?

3

ROLES AND RESPONSIBILITIES

The science department has existed in its current form for approximately 100 years. Over that time, the department has reflected the changing nature of the relationship between society and science. As science has acquired for itself greater prestige and power, so too has the science department become more entrenched at (or near) the top of the subject hierarchy found in so many secondary schools. This position has been reinforced by the close connections between university science faculties and departments and between disciplinary science and the academic script of science teaching. There is a tradition among science teachers as to what "good" science teaching looks like, and given how heavily teachers are socialized into this tradition, it is extremely difficult for an individual to challenge it alone. If, however, we believe that departments are places in which science teachers can begin to understand and challenge why they teach the way they do, and the imperatives for change, then we must also understand the roles and responsibilities of the person charged with the administrative management and instructional leadership of the department: the chair.

In this chapter, we start by considering how the role and responsibilities of the chair have evolved over the past 170 years. Following this history lesson, we will move on to consider the work of Jeremy Peacock who, working from the literature on science chairs, has highlighted four important leadership capabilities for contemporary science chairs looking to enact instructional leadership practices in their department. Those capabilities are then brought together with leadership theory to explore the relationship between departmental and instructional leadership. Establishing the links is not the same as providing a checklist that says "do these things and all will be well." It is a guide for understanding the nuances of leadership within the department. The hard work, as always, is to put the guide to the test in the day-to-day life of the department. Next, we will ponder the implications of the dominant current department structures on the leadership of the chair, before moving on to consider how chairs position themselves between the work of the department and the (often contradictory) requirements of districts and legislators. Finally, we will turn our attention to getting started on the road to reimagining the department.

The Chair: A Short History

The position of the chair has never been clearly defined, despite its key role in shaping instructional leadership within the department. The role has been seen at times as simply administrative: making sure that school policies are enacted and adhered to; at other times the chair has been tasked with ensuring that the examination requirements of the universities are met; and at still other times chairs have been given the responsibility for improving teaching and learning. Increasingly, however, all of these roles are being simultaneously delegated to the chair. One thing that has remained constant, however, is that the position has always been somewhat ambiguous, with little agreement on the functions or selection criteria. The more things change, the more they stay the same—since at least the 1840s.

Early Days: 1840s–1905

In the 1840s the early science educator Richard Dawes believed that the primary role of the teacher was to make "children observant and reflective; to make them think and reason about the objects about them … to instruct them in the school of surrounding nature, and to bring their minds to bear on the every-day work of life" (Layton 1973, p. 42). To achieve this, Dawes instructed the teachers in his parish schools in both content and how his curriculum was to be implemented. Dawes had little time for discussions into differentiated curriculum for different social classes. For him, teaching was a matter for which "the real difficulty of the question is not with the people, or the classes to be educated … but in getting it out of the hands of talking men and into those of the practical and working ones" (cited in Layton 1973, p. 48). The professionalization of science was to change this perception of the learning required by science teachers.

The establishment of science subjects that were closely aligned with the university disciplines had a profound effect on teaching and learning. For example, science (in the form of systematic botany) was established as a subject at the Rugby School in the 1850s and was taught as a "pure" science. Science was seen as a commonsense activity that required the learning of specific content and the laboratory skills needed to enter university science. As such, there was little effort to develop the pedagogical skill of the teachers. The role of the science chair was principally administrative, ensuring that the university-imposed standards were met. As we have seen, Michael Faraday spoke against the manner in which science was being taught, arguing that the result of an abstract scientific education was that even the supposedly well-educated were, in science, "ignorant of their ignorance" (Public Schools Commission 1864, p. 381).

Establishing Departments: 1905–1950s

From Kilpatrick's usage of the term *department* in 1905 until the middle of the 20th century, two important forces acted to shift the role of the chair away from the administrative focus of the early period. First, in the United States there was a significant growth of secondary enrollments driven by a number of factors: major demographic changes with large increases in immigration, the increasing urbanization of the population, and major changes in child labor laws. According to Sheppard and Robbins (2007), there was "an approximate doubling of the high school population every 10 years from 1890 to 1930" (p. 201). This increase was matched by the loss of influence of the "mental training" view of education. Science teachers began to assert themselves as more than scientists: They were also educators. Writing specifically on biology, Sheppard and Robbins (2007) state that:

> *There was a rejection of the college dominance of the biological sciences as being abstract and impractical ... High school teachers wrote the new biology texts, and the biology syllabi were adapted to the developmental needs of students who would be in the earlier grades. The content of the course was more practical. (p. 201)*

For chairs in the early 20th century, this meant the evolution of an increasing responsibility for pedagogy, supervision, and administration. The situation, however, remained quite fluid as the trend toward teachers' disciplinary education produced departments staffed by specialists who reinforced the academic script of science education. Unsurprisingly, the first empirical studies into the role of the chair concluded that the position was in a state of confusion, with little agreement on either the chairs' function or the criteria for selecting chairs (Peacock 2014). Later researchers reported that that the sources of this confusion were not dealt with. Chairs were too busy with teaching and administrative trivia to focus on their main function of instructional supervision, and many chairs were not consulted on personnel issues affecting their team of teachers. In 1947, Lowry Axley compared the role of the chair to that of a racehorse burdened with the duties of a plow horse:

> *The departmental plan is based on specialization, but apparently very few systems make full use of the specialized training of heads of departments. The owner of a champion racehorse expects a championship*

> *performance when his horse is put to the test, and he would be considered a congenital idiot if he burdened his racer with the duties of a plow or draft horse in addition to his racing ... Their main function is lost sight of, and they are not given proper opportunities to use their training to promote the efficiency of their schools. (Axley 1947, p. 274)*

In the 1950s, research began to focus on the potential importance of the chair to improving the quality of instruction within the department. Rinker (1950) suggested that chairs should maintain a simultaneous focus on supporting students and teachers, while developing links to academic, professional, and school communities, while also performing clerical duties. This focus has continued to be developed over the past half century.

Latter Days: 1960s to the Present

In the 1960s, changes in research methodologies allowed researchers to investigate the chair's work and to analyze the relationships between the specific factors that affect that work. These methodologies developed even as the publication of Schwab's "The Teaching of Science as Enquiry" touched off an ongoing questioning about the meaningful purposes of science education. Given that the pressures for reform are only intensifying, the capacity to differentiate between aspects of the chair's work is an important step in understanding the role and the impact that it can have on teacher professional learning. While the earlier concerns about the role of the chair continue to be reiterated, there is an increasing awareness that "chairs are in an ideal position to facilitate instructional improvement because of their daily contact with teachers and their own instructional expertise" (Weller 2001, p. 74). This recognition is based on a number of factors. As science teachers are socialized into their departments, chairs are in a strong position to offer leadership around teaching and learning. Consequently, departments can represent an important site for professional learning and also function as a link between teachers and other science education organizations such as the National Science Teachers Association (NSTA), the National Science Education Leadership Association (NSELA), and university science education faculty. The NSTA position statement "Leadership in Science Education" outlines the roles that science leaders, including chairs, have in the implementation of reforms such as the *Next Generation Science Standards* (*NGSS*). Unfortunately, despite the growing awareness of the potential for chairs to provide leadership, they remain underused as a resource for improving instruction (Weller 2001). The overwhelming picture remains of chairs being asked to do too much with too little for too long—of racehorses continuing to be being burdened with the duties of the plow horse.

So, what does current research tell us about the leadership required of chairs in implementing reforms such as the *NGSS*? The recent work of Jeremy Peacock, a former science chair and a regional content specialist from Georgia, highlights four important leadership capabilities for contemporary science chairs who are seeking to provide instructional leadership in their departments.

Leadership Capabilities

Peacock worked through the research literature on the roles and responsibilities of the science chair from 1910 to 2013. This material has been analyzed using the concept of leadership capabilities that can be defined as the "seamless and dynamic integration of knowledge, skills, and personal qualities … [required for a] practical endeavor such as school leadership" (Robinson 2010, p. 3). From this work, four core leadership capabilities emerged as contributing to the ability of science chairs to offer science instructional leadership:

- Science leadership content knowledge
- Advocating for science and science education
- Building a collegial learning environment
- Negotiating context and solving problems

The relationship between leadership capabilities and instructional leadership is shown in Figure 3.1 (p. 52). Peacock makes the point that, while the leadership capabilities are interdependent and carry equal importance, the particular arrangement of the capabilities is intentional. Given that the role of subject-specific leadership is generally underrepresented in the literature, science leadership content knowledge is given prominence at the top of the figure.

The value of Peacock's work is that it draws from the literature to provide a guide to the capabilities that chairs need to work with if they are to reimagine the department. Our advice to chairs regarding these capabilities comes with two caveats. The first is that science leadership content knowledge, while underrepresented in the literature, is critical if a chair is to establish credibility for any reform proposals. One of the major issues that plagues the implementation of many reforms is that they appear disconnected from the work of teachers. Teachers place great store in credibility, and the best way to build support for any reform is to allow teachers to see the reform in practice. The second issue is that we, as science teachers, will never possess all knowledge in these areas, nor should we be expected to. If we are to reimagine the department, we need to be aware of our

Figure 3.1

CONCEPTUAL MODEL OF LEADERSHIP CAPABILITIES CONTRIBUTING TO SCIENCE INSTRUCTIONAL LEADERSHIP

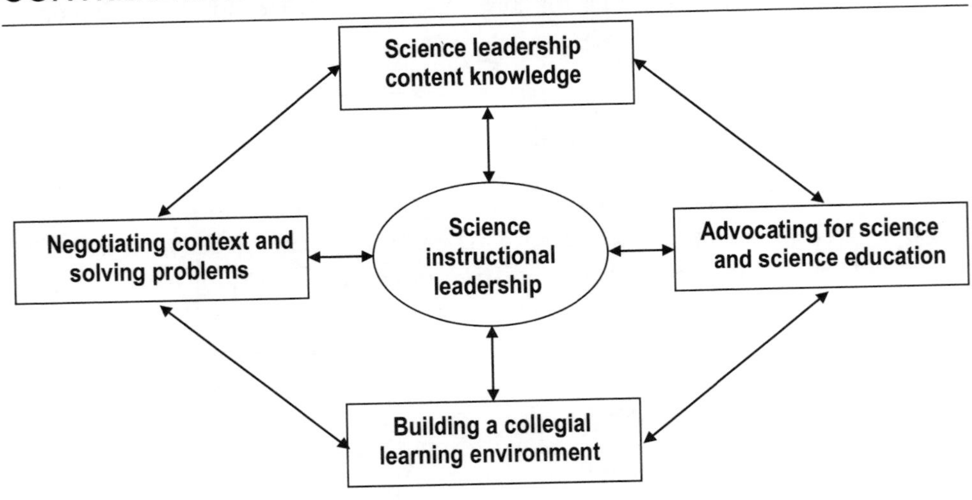

Source: Peacock 2014, p. 43

strengths and limitations and work from where we are. Paralysis through analysis serves no one, least of all our students and colleagues. We learn by doing and (hopefully) from our mistakes, so these capabilities evolve over time, reflecting changes in our own knowledge, the impact of mandated changes, and the changes that occur in departments as teachers also learn. The important point is that the capabilities focus us on what is important in different yet interconnected aspects of our work as chairs. So, let's take a closer look at each of the capabilities.

Science Leadership Content Knowledge

It should be obvious that a chair possesses a comprehensive understanding of science, but in saying that we open up an important issue that is sometimes ignored. Reform documents such as the *NGSS* are clear that discrete knowledge of science concepts is no longer sufficient. Teachers must be more than content specialists—they should *also* be learned generalists with the capacity to link science to the world in which they and their students live. The NSTA position statements also increasingly reflect this growing change in emphasis. For example, read the statement "Quality Science Education and 21st-Century Skills" (NSTA 2011). This is a call to recognize and value the personal practical knowledge that all teachers bring to their work. Personal practical knowledge, derived from experience within both the profession and general life experiences, is foundational to teaching.

For chairs, this leadership capability centers on three factors. First, they should possess, and be constantly refining, their reform-based expertise in science content, the teaching and learning of science, instructional strategies, curriculum, and assessment. As the new NSTA position statement on the *NGSS* makes clear,

> *implementing the NGSS requires that experienced teachers make a significant shift in the content and manner in which they have been teaching and that beginning teachers make a shift from how they were taught at the university level. For many teachers a modification in the content knowledge and competencies will need to be made.* (NSTA 2013)

This expertise builds credibility and allows other teachers to see what reform-based instruction looks like. Second, establishing credibility allows the chair to start influencing departmental curriculum and instructional and assessment decisions. This influence arises as the chair begins to facilitate reform-based learning opportunities for their teachers in areas such as instruction, curriculum, assessment, and student learning. Finally, if a chair is developing this capability, then he or she is in a better position to discern what is an educational fad versus what changes need to be made to improve student learning in every classroom.

Advocating for Science and Science Education

The links between departments and faculties of science have had a great influence on the historical development of departments. In 1950 Rinker expanded on these connections, suggesting that chairs should develop and maintain links to science in the wider community and be prepared to act as advocates for science. This is an important capability for three reasons. The first is that the development of links between the department and the wider scientific community opens opportunities for students to see science as occurring beyond the classroom. The NSTA position statement "Learning Science in Informal Environments" states:

> *The learning experiences delivered by parents, friends, and educators in informal environments can spark student interest in science and provide opportunities to broaden and deepen students' engagement; reinforce scientific concepts and practices introduced during the school day; and promote an appreciation for and interest in the pursuit of science in school and in daily life.* (NSTA 2012)

Secondly, chairs can advocate for science, the teaching and learning of science, and increased public understanding of scientific concepts. This is particularly important when confronting issues that may be contentious within society but not within the scientific community. The actions of the Dover Area School District biology teachers who refused to read the statement that "Because Darwin's theory is a theory …" in 2004 is one of the more extreme examples of teachers having to advocate for their discipline and profession.

Finally, to lead reform, chairs need to be actively engaged with developments in science education. Engagement is vital because it provides a frame of reference to gauge the position and performance of the department relative to what is being mandated by the reform documents. The relationship between science, science education, and society has changed. Without an understanding of those changes, and an awareness of the alternatives to what currently happens in their departments, chairs may not be able to see beyond their concerns with the covering of the curriculum to the wider issues that they should be addressing.

Building a Collegial Learning Environment

A collegial learning environment is far more than a place where teachers enjoy the company of their colleagues. Chairs have the key role in shaping departments as places where teachers share a responsibility for the continuous improvement of student achievement. Such an environment has three characteristics, the first of which is the need for a focus on both teachers' content and pedagogical knowledge and students' ways of learning content. Second, there must be opportunities for teachers to engage in active learning through activities such as mutual observation and critique, the collaborative implementation of innovations, and opportunities to review student work and assessment and communicate these to other teachers. Third, learning opportunities need to be coherent with what teachers already know and work from that point to move toward the ideals of reform documents such as the *NGSS*. A collegial learning environment is one in which teachers are prepared to learn how to analyze their own and each other's instructional strategies, consider the links between teaching and learning, and experiment with alternative instructional and assessment strategies. To develop a department along these lines requires the chair to take a leading role in modeling these qualities while also being aware of the context in which their department operates.

Negotiating Context and Problem Solving

Chairs are impacted by a range of sociopolitical forces (and their attendant values) including those that operate within the school, the national education policies, and the complex range of forces that are conveniently grouped under the banner of globalization. All of these have some effect on the work of the department. The challenge for the chair is to negotiate through these forces and simultaneously work to improve teaching and learning and meet the demands of policy. This is never an easy task, and there are times when the wrong decision will be made. The best advice that we can offer is to work with your department to make the most morally defensible decision that can be made in support of any movement toward the ideals of the reform documents. Sergiovanni (1992) has described five bases from which leaders can draw their power: bureaucratic power, based on rules and regulations; technical-rational power, based on the leaders' knowledge of the field; psychological power, based on knowledge of human relations; professional power, based on professional norms and standards; and moral power, based on clearly enunciated values and the shared norms of a community. In part, leadership involves making political decisions and having the power to carry them through, and the chair who maintains a moral presence is more likely to shape the department as a community committed to improving teaching and learning and reimagining the department toward the ideals of the reform documents.

Peacock's model gives us an understanding of the capabilities that chairs need to bring to, and continue to develop in, their role. Leadership is about people, and the capabilities that we have discussed here all contribute to developing the conditions that allow departments to act as places for teachers' long-term professional learning. How those capabilities can be brought together as a coherent whole is the focus of the next section.

The Department and Leadership

One of the key principles that directed the writing of this book is the belief that individual science teachers struggle to align their work with the practices outlined in documents such as the *NGSS*. In saying this, we are not questioning the commitment of any science teacher; rather, we understand that the academic traditions the overwhelming majority of us have been socialized into make it profoundly difficult for us as individuals to make the sorts of changes to instruction envisaged by the reform documents. We also believe that departments, as both communities and organizations, possess many of the qualities needed to support the professional learning of all science teachers, and that the role of the science chair is crucial in realizing that potential.

Let us be clear, professional learning is more than the acquisition of knowledge; it is a preparedness to question current pedagogy and develop new instructional strategies that improve the quality of teaching and learning. Yager (2005) states that "the focus of teachers' learning should be one of inquiry into teaching and learning. This, of course, emphasizes the use of questions that leads to learning and the identification of possible answers" (p. 17). At its heart, therefore, the importance of the department to professional learning lies in its capacity to develop as a trusting environment in which teachers are free to question the teaching and learning of science. Such an environment provides

> *opportunities to voice and share doubts and frustrations as well as successes and exemplars. They need to ask questions about their own teaching and their colleagues' teaching. They need to recognize that these questions and how they and their colleagues go about raising them, addressing them, and on occasion even answering them constitute the major focus of professional [development]. (Lord 1994, p. 183)*

This quote raises two important questions: What does this environment look like, and is there some process to facilitate opportunities for learning? To answer these questions, we return to our conceptualization of the department as being a community and organization simultaneously. As we saw in Chapter 1, this dual conceptualization enables the chair to choose the appropriate cultural or bureaucratic strategies, or some combination of both, to pursue the aim of reimagining the department.

As a community, the department has a primary role in shaping teachers' instruction. This does not imply that all teachers in the department will share identical images of science or science education (see Wildy and Wallace 2004). However, as science teachers, they all share a particular identification with the discipline and subject. It is this common identification that serves as the starting point for conversations into the teaching and learning of science. The aim of these conversations should be to develop a consensus of what is important in science education and from there establish clear goals for teaching and learning. It is the development and communication of these goals that becomes the source of political power of the department as an organization.

In conducting conversations around the teaching and learning of science, remember that resources such as the *NGSS* and the NSTA position statements, work with board and state or provincial science specialists, and attendance at conferences can all provide valuable insights and supports for teachers and chairs looking to make changes to their instructional and assessment strategies. There is no justification for reinventing the wheel. Without input from outside the department, there is a real risk that conversations can be used to reinforce the status quo. Therefore, the chair must be prepared to use the position to shape the conversation toward the goals of the reform documents and his or her vision for the department.

Transactional Leadership

The initial steps in shaping the department as a community involve understanding where teachers are in their professional (and to some extent in their personal) lives, their understanding of teaching and learning, and their learning needs. Initially, this may well involve the chair in self-interested exchanges with self-interested others and being prepared to bargain with teachers whose "interests and claims serve their own goals primarily, and only secondarily, if at all, serve the interests of the organization" (Starratt 1999, p. 26). Realize that there will always be some teachers who are set in their ways and reluctant to change, if they will change at all. As chair, you will win some, and you will lose some. Don't take it personally and never lose sight of the bigger picture.

The leadership literature refers to this as transactional leadership, and it is really about establishing the ground rules by which teachers can participate in the work of reform. This form of leadership is concerned with the bureaucratic issues of supervision and organization to promote "a routinised, non-creative but stable environment" (Silins 1994, p. 274). In establishing these ground rules, there must be a commitment to values such as integrity, honesty, trust, wisdom, and fairness and to the needs and rights of all involved. Central to these conversations must also be a sharing of instructional strategies and beliefs with other teachers and a constant message that the student success in learning and assessment tasks is the absolute priority in everything the department does. Setting the ground rules through transactional leadership is an important first step in shaping the community, but it will not by itself lead to long-term commitment, and there will be times when the chair will have to revisit the ground rules. Teachers come and teachers go, and as issues arise, the ground rules will need to be reset. The importance of transactional leadership is that it sets the stage for the department to move beyond being a collection of science teachers toward being a community of science teachers who are prepared to reconsider their instructional strategies in light of the reform documents. This brings us to what is known as transitional leadership.

Transitional and Transformational Leadership

To effect long-term change requires personal commitment. Teachers need to know that the chair is supportive of them and their work; the chair must establish a "moral presence" (Starratt 1999). Such a presence is grounded in how the chair works with his or her teachers and must reflect the virtues of honesty, courage, care, fairness, and practical wisdom. The conduct of the chair is crucial in building respect and a sense of loyalty, both of which are foundational to any movement beyond transactional leadership toward transitional and transformational leadership.

Transitional leadership moves beyond the stability of transactional leadership and begins to challenge the status quo of individual and departmental teaching and learning. It does this by beginning to draw on the (perhaps latent) abilities of teachers to create new standards of expertise and collegiality, shared values and beliefs, and a shared commitment to the work of the department. It is this shared level of commitment that gives the department its political power. A strong department is one that is characterized by "individual and communal empowerment … involves the gradual embracing of responsibility for one's actions. It involves autonomous individuals in the choice to be active, rather than passive" (Starratt 1999, p. 29).

It is also one in which the difficult decisions that often have to be made about teaching and learning and issues such as resource allocations can be made in an informed way. The conduct of the chair continues to be crucial at this stage for a number of reasons. Teachers need to be able to trust in the chair that it is acceptable to make and learn from mistakes. The realignment of relationships (e.g., from individualistic teacher to colleague) and the professional conversations that underpin that realignment must be based on honesty and care. For teachers to move beyond the pedagogies that have served them well in the past and to embrace new pedagogies is an act of courage. It is also a stage that is risky and cannot be rushed. Starratt (1999) suggests that the transitional stage may take two to four years. In our experience and working with other chairs, this time frame may be on the optimistic side. Sustained over a period of time, transitional leadership can take on aspects of transformational leadership, which

> *seeks to unite people in the pursuit of communal interests. Motivating such collective action are large values such as community, excellence, equity, social justice, brotherhood, freedom. Transformational leaders often call attention to the basic values that underly the goals of the organization, or point to the value-laden relationships between the organization and the society it serves. Transforming leadership attempts to elevate members' self-centered attitudes, values and beliefs to higher, altruistic attitudes, values and beliefs. (Starratt 1999, p. 26)*

Given that one of the main objectives of reform documents such as the *NGSS* is to overcome the growing distance between an academic science education and contemporary students, we believe that transformative leadership at the departmental level is crucial. Only when teachers understand the need to change, are presented with viable options for change, and work in an environment where they can safely take on an increasing personal responsibility for shaping and living the values and interests of reforms will change be sustained. We would caution, however, that the shift from transactional to transitional and then transformative leadership is never linear; there will be movement backward and forward. Departments are never static, so the chair must constantly be proactive in developing opportunities for teachers to revisit and renegotiate what is important to them as a department in light of their growing expertise in working with the reform documents. This requires knowledge of the predominant departmental structures that exist for contemporary science departments and how those structures can influence leadership.

Departmental Structure

Busher and Harris (1999) have investigated the structure of departments and the implications of those structures for the leadership of the chair. Of the five structures they describe, two are of particular importance to science. Both structures are characterized by having a number of teachers and access to a range of resources. The first structure, the "unitary" department, is more likely to be found in smaller secondary schools in which there is a limited differentiation of science into its component areas. In such departments, the chair can exercise a strong and direct influence on the teaching and learning that occurs. The other structure, the "federal" department, is more likely to be found in larger secondary schools and may consist of specialized subject leaders tasked with shaping teaching and learning for their specialization under the aegis of the department. In a federal department, the chair needs to supervise and coordinate the work of these specialist leaders within the framework of the whole department. Federal departments generally work because "their subjects and pedagogies are perceived as cognate and their cultures are substantially homogeneous" (Busher and Harris 1999, p. 309). The different structures, however, clearly place different leadership demands on the federal chair compared with their unitary counterpart.

Unitary departments, given their generally smaller size, require leadership from the chair that balances and prioritizes the needs of various courses within the science program. This requires a level of skill in dealing with political demands for resources and developing the formal and informal strategies for coordinating teaching and learning with the needs of students. In contrast, federal departments require leadership at both the specialization and department level. Important factors to consider with these departments include the history of the department's development and the consequent impacts on the formal and informal distribution of both power and authority. It is not difficult to visualize a long-serving teacher in an area both possessing and being willing to use their influence to protect or promote the interests of their specialization. In an earlier work, one author of this book reported on a chair who acted as a mentor for a teacher who subsequently became a chair:

> *Will related that the teaching of school science in the mid-to-late 1960s was rigidly organized into scientific sub-disciplines: "I was a biology guy, and my first departmental chair was a physics guy—physics guys and biology guys don't think the same way, and we don't pretend to." This siloing of knowledge was obvious to Dan when he started teaching in 1982: "Teachers guarded their territories within science. So a physicist was a physicist, a chemist was a chemist. And we had grade nine and ten courses that you had to teach, but you always taught it from your perspective. It was a big competition, I think, to see whose science was best and where the kids ended up. Which science will they pick?" (Melville and Bartley 2010, p. 812)*

If working in a federal department an important point to note is the extent to which the chair, at the center of the department has, and is recognized to have, sufficient power to lead. Without power to effect change, leadership will not happen. And that brings us to the final section of this chapter, the relationship between chairs and those external decision makers whose demands so often impact the work of the department.

The Chair and External Forces

As we all know, reforms in education often appear to come and go, and the more cynical among us believe that the more things change, the more they stay the same. Science education has not been immune to this, and reform efforts have attempted to respond to the growing disconnect between science, science education, and society by explicitly outlining how teachers and their classrooms need to change. Given the resilience of the academic tradition in science, we could argue that documents such as the *National Science Education Standards* and the *NGSS* have been very good at saying *what* should be taught, yet have fallen short in understanding the *how* of enacting and supporting these reforms. This is particularly important for chairs because they are often under pressure to simultaneously provide instructional leadership in the department while implementing a range of curriculum and administrative changes. These changes can come from the school, board, state, and national level, and sometimes appear contradictory to the teaching and learning of science. How the chair responds to (or positions him- or herself in relation to) these external forces is crucial if the department is to maintain a focus on improving the teaching and learning of science. To consider how chairs can position themselves in relation to reform initiatives, we turn to the work of the French sociologist Pierre Bourdieu.

The position of the chair to reforms is influenced by both their personal dispositions to the reforms and what the department as a community and organization sees as good science teaching. For most departments, this is a continuing attachment to the traditional script. The strong personal and professional relationships found in departments allow them to be considered social spaces or "fields" (Bourdieu 1990). Fields can be conceptualized as specific social environments "with explicit and specific rules, strictly delimited in … time and space" (Bourdieu 1990, p. 67). As social constructs, fields are made up of individuals who share common beliefs and practices and compete for symbolic and material products or "capitals" (Bourdieu 1984). Extending the concept, schools, boards, and reforms such as the *NGSS* can all be seen as fields with their own particular rules, priorities, and values. These fields are never independent of each other—they all overlap and exert an influence on teachers and classroom teaching and learning. Fields come into conflict and competition when what is valuable to each is challenged. If a reform is seen to challenge the instructional strategies that are valued by a department (and are seen as foundational to the department's power and prestige), then the reforms will be resisted. That resistance can take many forms, from rejection to co-opting the reform to the values of the department. Alternatively, a reform that is seen to reinforce the values of the department will be accepted. To a large degree, the response of the department to reforms relies on the leadership of the chair and his or her ability to understand the relationships between the "power structures, hierarchies of influence, and … practice" of both the department and the reform (Lingard and Christie 2003, p. 320).

In working with 12 science chairs in the southeast United States, Peacock (2013) identified two major constraints on chairs as they sought to understand and implement external reform efforts. These are important to understand in terms of highlighting the pressures that chairs face and the courses of action that are possible when dealing with reforms. The first constraint was how a chair's school context shaped his or her capacity to act as an instructional leader. Specifically, their position within the school leadership hierarchy constrained their leadership. Four types of leadership were identified: the chair as liaison, informal shared leadership, formal shared leadership, or the chair as autonomous leader. At one end of the hierarchy, the chair as liaison implements school (or board) administrative initiatives within their departments. In the second group, chairs who have negotiated greater authority exert a more active influence on instructional practices. The third group possesses a formal leadership position that gives direct access to school-level decision making. The final group enjoys some level of autonomy from school-level administration in shaping the direction of their department. In presenting these groups, we are aware that the chair's position within the hierarchy of the school is not fixed, since changes within senior administration, staffing, and the changing issues that a chair faces can (and do) affect relative positions in the hierarchy and the responses that are required of them.

The second constraint was the influence of general education reforms on science chairs. Rather than seek to engage with science reforms, chairs were spending their time and effort addressing questions of assessment, accountability, and school improvement. Consequently, they were limited in the leadership they could provide in support of science education reforms.

We can learn from the experiences of the chairs in this study through a consideration of their words. To do this, we would like to offer a short vignette (drawn from Peacock 2013) from each of the leadership approaches we have described in this section and how they attempted to connect the work of their department to the wider reform efforts in science education.

Brad: The Chair as Liaison

Brad saw his role of chair as a liaison whose principal duty was to "push the administrator's agenda" within his department:

> *There was a big push in the district for teaching science with inquiry. I was pretty excited about the possibilities [and] initiated a program in which a group of teachers would go through the curriculum and identify specific ways to infuse inquiry-based strategies into the district curriculum. The group worked very hard trying new things, planning activities, and discussing outcomes. Ultimately, it all fizzled. Administrators and teachers ultimately didn't buy into the effort. I have come to believe that good standardized test scores are really all that matters to the bureaucracy. If the test scores in the paper look good to the public, initiatives from the grassroots aren't going anywhere—even when they would be good for students.*

Consider Brad's position for a moment. He was excited about the teaching of science as inquiry and had the freedom to initiate a program that encouraged teachers to try new strategies. And yet, in his own words, the effort ultimately "fizzled." But why? Have you, or your department, ever been in the position to enact a district agenda and given the resources to do it? What happened, and more importantly, why did it happen? What is the responsibility of the chair in such a situation? We would suggest that, when implementing a district (or state) agenda, how you view the role of the chair is crucial. In this vignette, Brad saw his role as implementing the district agenda rather than as an instructional leader. Consequently, his understanding of the reforms and how they could be translated to the classroom was limited. While Brad viewed the inquiry initiative positively, he had not instituted the change and did not display a long-term commitment to transforming science teaching. Second, Brad appeared to lack both the time and influence to challenge the teachers' and administrators' perceptions of the reform. To challenge the status quo requires power and influence to be wielded for considerable lengths of time: a one-year appointment is not credible. Further, the strategy that Brad implemented indicates a limited capacity to integrate the work of the department and the reforms. Without a strong understanding of the reforms, the decision to "go through the curriculum and identify specific ways to infuse inquiry" indicates an oversimplification of the complexity of inquiry and teacher professional learning. Similar concerns will arise with the *NGSS* and its emphasis on practices. What can you learn, or need to learn, from Brad in terms of your understanding of reforms and how to introduce them to your department?

Charles: Informal Shared Leadership

Charles was in the position of, having been elected to the position of chair, also being responsible for conducting the school teacher performance evaluation process. Within his school, elected chairs experienced little support from school administrators. The teacher performance process was seen as poorly designed, with positive ratings being perceived as doing well due to the teacher's efforts, while negative ratings were perceived as punitive and not the responsibility of the teacher. The net result was that long-term professional learning was not encouraged. Consequently, Charles saw the role of chair as intensely political, balanced between influencing and alienating the teachers in his department. Charles also reported that his main concerns as chair were laboratory and chemical safety and student participation in science fair competitions. Important as these are, they are not reform issues. For Charles, the implementation of the *NGSS* was reduced to a question of content: "We will need to rework our curriculum maps …" Charles did discuss several examples of instructional leadership within his department. In particular, he attempted to introduce teachers to a series of board-mandated content literacy strategies:

> *I'm going to target the ones who are struggling. Now, you have to be very subtle because I don't have any ability to make anybody do anything. I can give [poor ratings] now and then, but it's just punishment; the rating system is not designed very well. The teachers will just say, "We*

> *don't want you anymore." As I've been trained in the reading across the content area, I can take some of those literacy strategies and say, "Hey, let me show you this." I'll just show it to them, and then they try it, and they'll talk about and say, "Well, this was the problem." What I'm trying to do is repair the places that I think need repair, whereas they will try the strategy and then forget about it.*

Charles believed that his work as chair was limited by the "top-down approach from the central office" in which administrators directed a series of general literacy and assessment initiatives. Charles' position is more common than we would like to admit and is a major source of stress and frustration. As a reality for many chairs, situations like these raise several issues. Before reading any further, what issues does Charles' dialogue raise for you? What are the leadership capabilities that need to be evident (or developed) in a situation like this? To what extent is Charles's perceived lack of influence indicative of a greater need to understand the department as a community? By this we mean that the chair needs to understand where teachers are in their professional (and possibly personal) lives, their understanding of teaching and learning, and their learning needs. Effective instructional leadership is based on an understanding of people, both teachers and students, and their learning needs.

Kim: Formal Shared Leadership

As a department chair, Kim occupied a formal position in the school leadership, with access to school-level decision-making processes. At the time of the study, the administration was focused on the use of student assessment data as a basis for decision making. Consequently, the focus of her work was closely aligned to the goals of the school, not the department. Her departmental leadership was evident in the operation of a STEM (science, technology, engineering, and mathematics) academy within the school and working with her district science coordinator in providing active support for science teachers.

> *As part of the STEM academy, I have been working to increase inquiry-based learning, depth of content knowledge, and reading and writing across the curriculum. The purpose is to get students to develop a deeper understanding of content material and to be able to communicate and apply those ideas to other areas. The* NGSS *will definitely add to the supporting framework to help all teachers improve mastery of standards, even if they are not in the labeled STEM courses. We will use the* NGSS *standards to provide an additional framework in conjunction with the* Common Core *standards.*

For chairs, who operate in the middle management of the school, this can raise a question of loyalties: to whom do you owe your loyalty, the school or the department? This question is fraught with danger, as Kim's vignette demonstrates. While Kim appeared to exhibit an understanding of science education reforms, including STEM and inquiry-based learning, there was a discrepancy between her words and the future of science education described by the reform documents. The first discrepancy was her approach to the reforms, which could best be described as mechanistic: "We will use the *NGSS* standards to provide …" The *NGSS* were portrayed as a checklist to prepare students to meet the *Common Core State Standards*, not as a long-term strategy for improving the teaching and learning of science. The second discrepancy was the unwillingness to challenge the academic script: The administrative focus was on assessment- and data-based instructional interventions, not on science education reform. Consequently, Kim mounted little challenge to tightly held beliefs, and the department remained on the periphery of science education reforms.

Fortunately, loyalty is not a zero-sum game, but it does require that chairs ask questions of themselves, their departments, and administrators. These can include questions such as the following:

- How can reform in the science department align with the aims and goals of the school?
- How do we understand student success in science, and how does that translate across the school?
- How can the professional learning opportunities in the department meet school-level professional learning objectives?

Melanie: The Chair as Autonomous Leader

Melanie was the chair of a department that had considerable freedom to chart its own course: "Nobody gets in our way much." With the support of her administrators, Melanie gave the teachers in the department the authority to pursue their own agendas. One outcome of this was the formation of a math–science academy within the school. Melanie herself took the primary lead for the physical science courses in her department, while relying on another teacher to lead the life science courses.

> *I went to one of the STEM programs that help with resources, and I started a robotics team because I'm trying to get an engineering design course this year. I'll be teaching that and going to camp with some of my students. We're also going to learn engineering design processes. …*

> *The other teacher is doing something similar, but he wants to do more project-based teaching within biology, and so a few years ago we started a math–science academy. It is supposed to be a capstone project at the end, but we haven't been allowed the time for them to work on this.*

Melanie wielded considerable power in her efforts at instructional improvement, but her efforts did not represent a specific commitment to the reforms described in the *NGSS*. The changes within the department were somewhat superficial and lacked the coherence of the reform documents. Teachers, free to hold individual perceptions as to the meanings of the reforms, may implement changes consistent with the reform but are more likely to adapt the language of the reforms to their existing instructional strategies (Stigler and Hiebert 1999). To be a chair is to accept the responsibility for the teaching and learning of science within the department. This means that you need to make value judgments as to what is important, and then see these decisions through. There is no easy way around this. If you are to successfully reimagine the department, then *why* are the reforms important to you, your students, and your colleagues? Answering that question will help you shape a coherent response to the external forces that impinge on the work of all chairs.

Learning From These Chairs

What can we draw from the positions that these chairs held toward external forces? The first lesson is that chairs need to have a solid understanding of a reform before they attempt implementation. Although the *NGSS* documents recognize that it is the teachers' responsibility to enact professional autonomy, there is a focus on prioritizing the engagement of students with the science and engineering practices, disciplinary core ideas, and crosscutting concepts. To achieve this focus, it is necessary for the chair to understand and identify these essential framing principles of the *NGSS* so that as the department undertakes reforms (as with Melanie's math–science academy) the *NGSS* can serve as a compass and measure of success. Documents such as the *Educators Evaluating the Quality of Instructional Products (EQuIP) Rubric for Lessons and Units: Science* (NGSS Lead States 2014) can serve to help keep *NGSS* central to all design work and discussions around teaching and learning.

Without tools to help guide the work of reform, the risk of possessing a superficial knowledge is twofold: It will either not engender real commitment or will be misinterpreted and run the risk of becoming coopted into existing practice. Chairs need to have an understanding of how to wield power and position in the promotion of reform and have the ability to prioritize their efforts. They also need to have the capacity to operate simultaneously and strategically within, and across, both the department and the reform. This involves developing the leadership capabilities that we discussed earlier: science content leadership knowledge, advocating for science and science education, building a collegial learning environment, and negotiating the context and problem solving. It is easy to write these words; it is much harder to live them, especially when faced with competing reforms. In the next chapter we start the journey toward putting these words into practice.

Summary

- The role of the science chair has historically been ambiguous. There is something of a consensus that the role involves a simultaneous focus on clerical duties, supporting students and teachers; and cultivating links to the wider academic, professional, and school communities.
- Leadership capability requires an integration of knowledge, practical skills, and personal qualities.
- For science department chairs, capability is required in four areas:
    1. Science leadership content knowledge
    2. Advocating for science and science education
    3. Building a collegial learning environment
    4. Negotiating context and solving problems
- Teacher professional learning is more than the acquisition of knowledge. It is a preparedness to question current instruction and develop new instructional strategies that improve the quality of teaching and learning.
- Departmental leadership is iterative, never static or linear. Depending on the context, chairs initially need to engage in transactional leadership, which sets the ground rules by which teachers can participate in the work of reform. Bureaucratic issues of supervision and organization need to be worked through at this stage. More importantly, chairs must demonstrate a commitment to values and the individual's needs and rights. Such a commitment is demonstrated in the sharing of instructional strategies and beliefs, and a constant message that the learning of all students is the department's absolute priority.
- Transactional leadership will preserve the status quo. If reform is to occur, then chairs need to draw on the abilities of teachers to create new standards of expertise and collegiality, shared values and beliefs, and a shared

commitment to the work of the department. This is transitional leadership and may take three or more years of work.

- Transformational leadership will only occur when teachers understand the need to change, are presented with viable options for change, and work in an environment where they can safely take on an increasing personal responsibility for shaping—and living—the values and interests of the reforms for the benefit of their students.
- The shift from transactional to transitional and then transformative leadership is never linear. There will be movement backward and forward. Departments are never static, so the chair must constantly be proactive in developing opportunities for teachers to revisit and renegotiate what is important to them as a department.
- Chairs need to understand the structure of their department and how that affects the politics of their work. Unitary departments require skill in dealing with the political demands for resources, and the coordination of teaching and learning with the needs of students. Federal departments require leadership that considers these departments' history and the distribution of both power and authority.
- The chairs' position within the school's administrative structure can shape the work of the chair. Chairs can occupy positions such as liaison, informal shared leadership, formal shared leadership, and autonomous leader. These positions are not fixed; changes in personnel and situation can affect the role of the chair.
- Chairs are also impacted by the time and effort required in response to general education reforms. These can limit the leadership that chairs could provide in support of science education reforms. Although chairs represent an important potential resource for supporting reforms, many chairs are seriously constrained in their ability to fulfill this potential.
- Chairs who are looking to reimagine their department need to have a solid understanding of a reform before they attempt implementation. Second, chairs need to have an understanding of how to wield power and position in the promotion of reform, and the ability to prioritize their efforts. And third, chairs must have the capacity to operate simultaneously and strategically within and across both the department and the reform.

Where Am I Today? Questions to Ask Yourself

For You as a Science Teacher

1. What does professional learning mean to me, and what responsibility do I take for my own learning?
2. How can I contribute to the professional learning of my colleagues?
3. What can I learn from my colleagues, and how can I establish those relationships?
4. In what ways might my actions around questions 2 and 3 help me to generate a culture of trust among department members?
5. What is my active involvement with professional associations such as NSTA?
6. What can I learn from the chairs whom I have worked with? What were their strengths and weaknesses? What would I do differently?

For You as a Department Chair

1. How do I see the role of the chair, and what do I really want to achieve over the coming year? the next three years? the longer term?
2. How do I prioritize my work as a chair?
3. What do I understand by the term leadership, and what do I need to learn?
4. What is the structure and history of the department? How do these influence the decisions that are made?
5. Who in the department possesses (and uses) power and authority? To what end is that power used?
6. What do I understand moral presence to be, and how would I seek to establish it?
7. What external forces do I have some influence over, and what is beyond my influence?
8. In terms of the leadership capabilities, what do I already do well, and what evidence is there for this judgment?
9. What leadership capability should I initially focus on developing? What resources will I need to develop my expertise in this capability?

4

GETTING STARTED

So far we have considered the historical development of the science department and the role this history plays in replicating a traditional view of science education that centers on the primacy of conceptual knowledge. We have looked at how contemporary practices in teacher professional learning have generally failed to bring us closer to the ideals of science education described in the *National Science Education Standards* (NRC 1996) previously or the *Next Generation Science Standards* (*NGSS*) more recently. Therefore, the *NGSS* make it clear that implementing and supporting reforms that engage students in science and engineering practices, and the use of disciplinary core ideas and crosscutting concepts to explain phenomena or solve problems, will require changes in teachers' professional learning—changes that are intimately linked to the roles and responsibilities of the department chair.

These changes include working to align teachers' views of science education with the vision of the *NGSS* and their own classroom needs and strategies. Further, teachers will need to develop a rich understanding of the scientific ideas and practices they are expected to teach and be able to draw on a range of appropriate instructional strategies that support this new vision of the teaching, learning, and assessment. As we noted in Chapter 3, chairs must also be constantly aware of the pressure that school boards and administrators can place on teachers to focus on measurable outcomes. This pressure is often to the detriment of teachers' efforts to align their practice with professional standards for science teaching. Chairs, therefore, have a professional responsibility to promote and defend an environment that values the teaching and learning of science. There are times when the role of the chair is not for the faint of heart.

If we believe that a sustained change in science education has to be focused at the departmental level, then the question you are likely to ask is, "Well, where do I start?" Having read this far, you have already started to answer this question and, more importantly, begun to understand the conditions that will help promote the professional learning of both yourself and your departmental colleagues. In this chapter, we focus on the steps that a chair needs to take if they are seeking to lead and manage a department that is beginning to move toward teaching and learning as envisioned by the *NGSS*. Changes such as these require time, leadership, and a supportive environment—all of which can be found in the science department. We have already discussed the crucial roles and responsibilities of the department chair in promoting an environment that is supportive of a reformed script for science education.

Before we go any further, we need to be clear that reimagining your department will not be a short-term, linear process. Promoting a reformed script across a department is more than attempting to reorganize the content that is to be taught. It is about the long-term establishment of a culture in which teachers inquire together simultaneously as teachers and students of science. Taking a long-term view will be problematic in schools in which the role of the chair is voluntary or rotated among teachers within the department, and as is often the case, no time is allocated for the performance of the role. In such cases, remember that curriculum leadership is not limited to the formal position of chair; individual teachers have the capacity to move toward the teaching and learning delineated in the *NGSS*. Research shows us that these individuals are often seen as leaders who open professional learning opportunities for their colleagues. If you are in a school in which the role of chair is not associated with the long-term reimagining of teaching and learning, then our advice to you would be to accept the challenge, persevere, and act as a catalyst to develop a common departmental agenda that supports reform. Your students and colleagues will come to appreciate your efforts on their behalf.

The nurturing of teacher learning is a well-accepted precursor to the encouragement of active student learning. Metz (1986) observed that the culture found in classrooms is a reflection of the departmental culture. Departmental cultures are largely defined by teachers' understandings of the nature of science, and among your colleagues those understandings may run across a wide spectrum. The range of beliefs extends from naive realism, in which science knowledge is seen as an "objective" absolute truth mirroring reality and directly accessible to human senses, to a more sophisticated belief about science as a tentative and evolving truth constructed as human explanations of natural phenomena. This evolving truth is crafted over time through a reliance on multiple theories and coordinated with data that is derived from rigorous scientific inquiry. Given this range of potential beliefs, the teachers in your department may express (and model) different, and sometimes contradictory, goals for science education. Understanding this is critical because the departmental culture shapes how the department organizes content, assessments and evaluations, and the allocation of resources. More importantly, the culture also shapes what a "good" science education looks like, and the values that shape teaching and learning (Carlone 2003). To be honest, any attempt to impose your views on the department will most likely be resisted—and defeated. This is why we suggest you start in the place where you have the most influence: your classroom. From within your classroom, you can start to build the credibility for the reforms that you are proposing, and credibility is foundational to transactional leadership.

Starting the Conversation

Let's consider the situation that exists in the majority of science departments today. Teaching science that is rooted in engaging students in science and engineering practices to explain phenomena or solve problems continues to be more the exception than the rule. One of the major reasons for this can be found in the lack of clarity that surrounds the meanings traditionally attached to inquiry in standards documents prior to the *NGSS*. Although teachers generally believe that the vision articulated in the *NGSS* is worthwhile for science education, there is still uncertainty as to the extent science should be taught as content or process, or some hybrid of the two, especially since it is acknowledged that relatively few rich visions of three-dimensional learning are available in practice. Overlay this uncertainty with the pressing needs that science teachers feel to teach facts and concepts that are mandated in state curriculum or standards and prioritized on tests and the difficulties that teachers face are entirely understandable.

The combination of uncertainty and pressure to conform to a traditional view of science education has resulted in teachers' not teaching a reformed script. The reasons given to support this decision (which is often made subconsciously) include the perceived difficulty of teaching from a constructivist perspective, the added time and energy needed, and uncertainty about how to meet the expectations of the curriculum. Often this is a perceived, or self-inflicted, pressure based on the traditional script's primary concern for content. Support from colleagues, the perceived physical limitations of the classroom, the costs of apparatuses and consumables, a belief that safety will be compromised, the placing of material in the proper sequence, and the demand of preparing students for further study are also cited as concerns (Baker, Lang, and Lawson 2002). One major concern that teachers often express is their uncertainty about the capacity of students to engage with the levels of analysis, argumentation, and evaluation described in the *NGSS*. This uncertainty is understandable but should be seen in the context of how we currently educate the majority of our students. If we do not teach a reformed script and engage our students in scaffolded and repeated opportunities to practice these skills, is it realistic to expect our students to move beyond a desire to say, "Just tell me what I need to know to pass the test"? And, if we fail to teach a reformed script, the question arises as to whether we have taught our students science at all.

The most pervasive concern for teachers, however, is also the most difficult to address. We are not just teachers; we are science teachers, trained in science and confident in our ability to teach it. We tend to resist any change to our work because it can be seen as a challenge to our identity as science teachers: that of the knowledgeable expert. Tytler (2007) describes the issue succinctly:

> *Part of the reason for the persistence of status quo science relates to the strong discursive traditions subscribed to by teachers of science resulting from their enculturation during their own schooling and undergraduate studies. ... This culture is strongly represented in school science discursive practices, supported by resources such as textbooks, laboratories and their associated equipment, timetabling arrangements and by assessment and reporting traditions. Another aspect is the force of long habit of teachers who have developed effective ways of delivering canonical content, who may lack the knowledge, skills and perspectives required for the effective teaching of a different version of school science. (p. 18)*

A move toward teaching a reformed script requires the teachers in your department to develop a new identity. Identities cannot be imposed; they must be built up over time through engagement in discussions about teaching and learning, experimenting with new ideas and teaching strategies and analyzing both the successes and failures, and paying attention to the emotional needs of your colleagues. For many chairs, to be deliberate in developing these changes is a very different form of work compared with the administrative roles that are usually expected of them. Let us be clear, the best place to start this work is in your own classroom. The conversation starts with you and your professional practice.

Building Credibility

Being in the position of chair does not automatically confer credibility on any departmental moves you wish to make toward reforming the teaching and learning of science. A leader's credibility is not based on position alone, it is also based on acknowledgment of expertise, experience, and respect. As an instructional leader, your teachers will often look to you to establish the credibility of the reforms that you are proposing before they will commit to change of their own. Establishing credibility involves two parallel components: developing a level of expertise in the reforms and clarifying the purposes as to why your and your colleagues' teaching strategies need to align more with the reform documents.

Developing a level of expertise does not mean that you know everything about the reforms—such an expectation is unrealistic. What it does mean is that you begin to question your existing strategies and begin to experiment by making small changes. Questioning is an important first step, and there are numerous ways in which you can begin to generate questions. Talk to your colleagues both within your school and further afield, read professional publications (e.g., *The Science Teacher*, produced by the National Science Teachers Association [NSTA]), actively participate in the discussions available through science teacher professional associations (e.g., the online NSTA Learning Center and subject Listservs), and develop links with local scientists and university science educators. As the NSTA position statement on professional development makes clear, the most effective forms of professional learning are those that challenge "deeply held beliefs, knowledge, and habits of practice" (NSTA 2006).

There are two potential dangers to be aware of as you begin to ask questions of your teaching. First, it is easy to be overwhelmed by the scope of the questions that you can ask: assessment practices, lab management, argumentation, and helping students construct explanations; the list is potentially endless. You need to employ your professional judgment and ask yourself honestly, relative to the *NGSS*, "What aspects of my teaching require me to reflect, reframe, and then reconstruct?"

It does not matter if you do not like the answers, what is important is that you acknowledge that these aspects can be improved. The second danger is the ever-present risk of "paralysis by analysis," where you spend so much time and energy analyzing your classroom strategies that you never actually begin to make any changes. This danger is particularly evident when we do not like the answers to the questions that we ask of our practices. In these cases, we tend to procrastinate. To move past this is an act of courage, and if we are not prepared to act ourselves, then can we really expect others to question their own practices? Credibility is built on having the courage to make the hard decisions and then act on those decisions.

Questioning aspects of our work relative to the *NGSS* is an important first step, but action flowing from that questioning is imperative. Teachers change their practice by tinkering, the process through which a good idea develops into "something worth subjecting to more systematic validation" (Hargreaves 2000, p. 231).

By acting on questions of teaching and learning, teachers begin to clarify the purposes of the change. It is this clarification that gives depth, value, and personal meaning to the change. Understanding why you are improving your teaching is important because it allows for informed experimentation, the pursuit of new ideas, and the freedom to learn from mistakes by refining those ideas and strategies for teaching and learning. The net result is that you begin to develop a broader repertoire of instructional strategies and the understandings that sustain them in support of the teaching and learning of the reformed script. Becoming systematic in making changes within your own classroom integrates theory and practice and gives coherence to your professional growth. In your role as an instructional leader, these qualities contribute to making you a credible source and catalyst for change.

If the conversation starts with you, then it is credibility, which in turn is leveraged as capital, that allows you to begin to engage your colleagues in the departmental conversation. This conversation is multifaceted, emphasizing "discussion, teacher collaboration, active inquiry, cooperative learning, continuous assessment of student understanding, and use of student experience and local issues as vehicles for learning" (Yager 2005, p. 23). With such a conversation, it is clear that there is no one best form of organization that is applicable to all departments. Our advice is to consider a framework such as that provided in the NSTA position statement "Professional Development in Science Education" as you initiate the conversation. Start small. For an example of how one teacher began the conversation, see Vignette 3 (p. 81).

Starting Small

Even as you are working in your classroom to develop a level of expertise and understanding of the purposes for the teaching and learning of the reformed script, you need to be actively cultivating the personal and professional relationships that will ultimately determine the future work of teachers within the department. Never underestimate the influence of relationships in shaping the work of teachers. Both the NSTA position statements "Professional Development in Science Education" (NSTA 2006) and "Principles of Professionalism for Science Educators" stress the importance of teacher relationships in any work to improve science teaching. As the chair, you have a particular role and responsibility in modeling behaviors that contribute to the development of solid relationships. By solid relationships, we mean relationships that allow you and your colleagues to openly discuss and critique (in the nonpejorative sense) teaching beliefs, strategies, and knowledge. The most important of these behaviors is your personal willingness to accept feedback and be seen as working toward improving the teaching and learning that occurs in the department. These behaviors build two qualities that are necessary for teachers' learning: respect and trust. It is only when respect and trust are present within a department that it is possible for teachers to join the conversation and begin sharing personal practices (Hord 1997).

A very effective strategy to begin engaging other teachers is to use your classroom strategies as a starting point for the conversation. Within your classroom you can model reformed teaching and learning; retain and display (as appropriate) exemplars of student work and assessments; develop anecdotes of your strategies and student responses; ask trusted colleagues to critique your classes; and mentor teachers through the sharing of ideas by discussing common concerns and clarifying the purposes of science education as they relate to the *NGSS*. Our advice is to start the conversation slowly and informally—over coffee, in the hallway, and in the science office. Start with those of your colleagues with whom you are most comfortable, but don't neglect your other colleagues. Keep in mind, especially if you are new to the position and your department has more experienced teachers, that your ideas may not be well received at first. If this happens, don't despair: keep developing your expertise and spend time building personal relationships with the members of your department. Solid professional relationships are always built on sound personal relationships. The aim of your work is to start small and integrate the conversation into your everyday routines. In many cases, these actions are not more work, they are different ways of viewing the same work that chairs are already expected to undertake.

In schools with dedicated time for teacher collaboration, make sure that these times focus on teaching and learning and not on administrative matters. If your school does not recognize the importance of collaboration time, then consider establishing a time; make it an opportunity to discuss teaching and learning. Keep in mind that one way to build the long-term political support of administrators is to invite them to your meetings and let them see the quality of the conversation. In starting the conversation from you own expertise (and not seeking to impose practice) you are starting to redefine the department as a place for teachers to develop reformed teaching and learning. This will occur (over time) for two closely related reasons. The first is that a conversation centered on your credible work will start to chip away at the cynicism that teachers often feel toward reform and reformers: "always changing and yet staying the same" (Stigler and Hiebert 1999, p. 100). Closely related to this is the second reason: By making the conscious effort to see teachers as producers rather than consumers of knowledge about teaching and learning, you are laying the groundwork for transforming the learning conditions in the department (Yager 2005).

The Hard Work of Changing Perceptions

This chapter has concerned itself, so far, with helping you initiate a conversation around the reformed teaching and learning of science. We have discussed the need for you to develop a level of expertise and a working understanding of the purposes of science education as expressed in the *NGSS*. From this starting point, we then started to expand the conversation out to other members of the department, with the aim of beginning to transform the teacher learning and teaching that occurs in your department. Let us be clear about the task you are starting on: Transforming your department is emotionally draining, challenging, and time-consuming. It is this way because powerful forces have socialized the vast majority of us, since our first science lesson in elementary school, toward a particular view of science and science education. In his book *Learning Science*, White (1988) wrote that "to change the way people teach requires a change in their perception of the context in which they work" (p.114) For science teachers to change the perception of their work is a difficult task but not an impossible one. For it to happen requires leadership; time; commitment; a clear sense of purpose; and a willingness to challenge teachers' tightly held beliefs, practices, and knowledge. These are the reasons that we stressed the need for you to begin developing your level of expertise, your practical understanding of the purposes of science education, and the importance of relationships founded on trust and respect.

Much contemporary professional development perpetuates the traditional view of science education, one that is frequently at odds with the teaching and learning of science envisioned in the *NGSS*. Far too often, teachers are presented with reforms that are principally concerned with content and argue for change from a deficit view of teachers' work. As Yager (2005) states, "Too often reform and improvement is defined as new organization of materials for teachers to use" (p. 16). Curriculum reformers often view educational change as a mechanical process, devoid of input from the teachers who have the task of implementing the change and living with the outcome. According to this view, the most effective way for teaching and learning to be improved is for teachers to find out what "researchers and policymakers say should be done with or to students, and then do it" (Lampert 1985, p. 191). In science education, this has translated into an emphasis on teacher professional learning that stresses the transmission of knowledge and skills by lecture and reading, the separation of science and teaching knowledge, the separation of theory and practice, teacher learning in isolation, the taking of additional courses and workshops, and an over-reliance on one-off workshops with external experts (NRC 1996).

There are two reasons why such emphases fail to transform the work of teachers. The first is that lectures and similar forms of transmission provide information to the listener; they do not provide knowledge to the listener. We have all been to one-day workshops that left us feeling inspired to change our teaching, but after two weeks in front of our classes we were hard pressed to remember what we had learned. For the information to become useful, the listeners must take it and give it meaning in terms of their practice. This takes time and is rarely achieved by individual teachers working in isolation. The second reason the emphases do not transform the work of teachers is significant in terms of the conversation you are starting in your department. The sharing of information about teaching and learning must be made in concert with the capacity to reflect on thoughts, words, and actions, and a willingness to act on those reflections. It is through thinking and doing that teachers learn about teaching (Clandinin and Connelly 1995). This is the value of the conversation you are starting. Once you have developed credibility for your reformed teaching and learning in your classroom, you are in a position to work with your colleagues as they start to inquire into their own teaching. The conversation you are initiating will begin to change the emphasis of teacher learning toward that envisaged by the *National Science Education Standards* (NRC 1996) and the *NGSS*. These standards focus on teacher learning through investigation and inquiry, the integration of science and teaching knowledge, the integration of theory and practice, collegial and collaborative learning, long-term coherent planning, a variety of learning activities, and an appropriate mix of internal and external expertise.

By this stage you are probably thinking, "This is all well and good, but how long will it take to get the conversation started?" This is an entirely reasonable question to ask, and our honest answer is that we don't know. Each department is different: different in the individuals who constitute it, different in its history, different in how it has socialized its members, different in how it views science and science education, and different in how it is viewed by the wider community. For these reasons, there is no formulaic answer to the question. What we can offer is a ballpark figure: You are looking at three years of work to develop a level of expertise and a working understanding of the purposes of the *NGSS*; to develop the level of personal and professional relationships that allow conversations that confront teachers' beliefs, knowledge, and practice; and to begin to develop those conversations. The figure of three years is based on our reading of research on teacher professional learning, and the experiences of ourselves and our colleagues over an extended period of time and on two continents. Remember, the figure of three years is to reach a point where the learning conditions in your department are beginning to be transformed. After three years there will still be much work to do, as we shall talk about in Chapter 5.

Summary

- Implementing and supporting the teaching and learning rooted in engaging students in science and engineering practices to use disciplinary core ideas and crosscutting concepts to explain phenomena or solve problems outlined in the *NGSS* will require changes in teachers' professional learning—changes that are intimately linked to the roles and responsibilities of the department chair.

- To expect individual teachers to make these changes without support is unrealistic. Change requires time, leadership, and a supportive environment, all of which can be found in the science department.
- A move toward teaching a reformed script requires the teachers in your department to develop a new identity. Identities cannot be imposed; they must be built up over time through engagement in discussions about teaching and learning, experimenting with new ideas and teaching strategies, and analyzing both the successes and failures, and paying attention to the emotional needs of your colleagues.
- A leader's credibility is not based on position alone. It is also based on acknowledgment of expertise, experience, and respect. Establishing credibility involves two parallel components: developing a level of expertise in the reforms and clarifying the purposes as to why you, and your colleagues', teaching strategies need to align more with the reform documents.
- As the chair, you have a particular role, and responsibility, in modeling behaviors that contribute to the development of solid relationships with teachers/colleagues in your department. By solid relationships, we mean relationships that allow you and your colleagues to openly discuss and critique (in the nonpejorative sense) teaching beliefs, strategies, and knowledge.
- The sharing of information about teaching and learning must be made in concert with the capacity to reflect on thoughts, words, and actions, and a willingness to act on those reflections. It is through thinking and doing that teachers learn about teaching.
- Each department is different: different in the individuals who constitute it, different in its history, different in how it has socialized its members, different in how it views science and science education, and different in how it is viewed by the wider community. For these reasons, there is no formulaic answer to how long it will take or exactly what it will take to effect change, only strategies for getting conversations started that can lead to meaningful work in the department and desired changes.

Vignette 3

STARTING THE CONVERSATION

Will began his teaching career in 1966, at a time when discipline boundaries within school science were fixed. Will had graduated with a degree in biology, and his first chair was "a physics guy." As Will points out, "physics guys and biology guys do not think the same way, and we do not pretend to." Despite this, Will developed a strong personal and professional relationship with his chair. Over a period of six years, this relationship gave Will the confidence to begin challenging his own classroom teaching. The catalyst for starting this challenge was twofold: a moral belief that his teaching was not engaging his students and the belief that he was not teaching the practices of science.

> *The cookbook experiments never sounded like science. They had their place, but it was really an illustration. It was not a way of thinking about science. It was better to act because I was not happy with how things were going and wasn't giving the kids a fair break. I had to devise a way to make it work better. If someone says this is the way we do things, but it isn't working, why would you keep doing it?*

For Will, the challenge involved moving beyond his own lifelong experiences (and success) with the traditional practice of science education. In accepting this challenge, Will relied on his own professional reading and collaborations with teachers in local schools.

> *Initially I had no idea of inquiry, and my teaching in the beginning was very traditional My labs were cookbook labs, as they had been for me at high school.*
>
> *As I read "Invitations to Enquiry" (Schwab 1962) my eyes were opened to a different way of looking at things. Additionally, a colleague had been to a workshop and come back with a number of inquiry-style labs. These sounded more like science, and I started to say that my students should be taking a problem, working out how to solve that problem, doing the experiment, and writing it up. The first lab was to show the effect of temperature, pressure, and concentration on the rate of enzyme function.*
>
> *My students had never done this before, and I was feeling my way with it. What I felt was important, however, was that I was doing something that made the learning better.*

Will went on to become a science chair in the early 1980s, and in that position influenced the teaching and learning of a number of science teachers and current chairs. Since retiring in 1999, he has remained actively involved in assessing students' inquiry-based culminating activities and adult education.

Where Am I Today? Questions to Ask Yourself

For You as a Science Teacher

1. How closely do my teaching practices align with the *NGSS*?
2. Why is the teaching and learning that focuses on engaging students in science and engineering practices to use disciplinary core ideas and crosscutting concepts to explain phenomena or solve problems important to me? To my students?
3. What areas of expertise do I need to develop in my teaching?
4. What strategies am I using to develop my expertise? Are there other strategies I need to consider?
5. What do I fear most as I work to develop my teaching? How do I overcome those fears?
6. What instructional and assessment strategies might I develop in my own classroom that could be viewed as exemplary in terms of the *NGSS* and pedagogical research?
7. How can I use these strategies as a springboard to engage my department in professional conversations about student learning and success?

For You as a Department Chair

1. What opportunities already exist for teacher professional learning within the department? Do they promote a long-term coherent vision of science education?
2. What mix is there in terms of impediments to change, and opportunities to break inertia within the department?
3. How effectively are those opportunities being used? If they are not effective, what needs to change?
4. How would I describe my personal and professional relationships with my colleagues?
5. What personal qualities (e.g., trustworthiness, the courage of my convictions, empathy) do I need to develop and model in my personal and professional relationships?
6. What views do my colleagues have of science education, and how are they evolving?

5

BUILDING FOR THE LONG TERM

This final chapter examines what it means to nurture a reimagined science department in today's context. Deciding to reimagine the department and initiating conversations with your colleagues is an excellent start. To sustain the momentum and encourage teachers to take greater ownership of the reforms will, to a large extent, depend on your leadership capabilities. These capabilities, and increasingly those of individual teachers, will impact and ultimately shape what the department looks like in the future. These capabilities are central to the chair and teachers making the paradigm shift away from the ways that they currently think, plan, operate, and perform to how they could do these things.

Departments do not, however, work in isolation from the rest of the school. To reimagine the department is to also be active in developing strong political and practical relationships with school administrators. Without their support, change is difficult to initiate and even more difficult to sustain. The aim of reimagining the department is to develop a long-term culture that is simultaneously owned by the teachers and supported by school administrators.

So, our purpose in this chapter is to consider four important matters. The first is the issues that chairs and teachers face as they work together to reform their work and how those issues can be confronted and overcome. You and your teachers will be challenged: That is a certainty. The second matter is the need to work toward distributed leadership within the department as a way of sharing the workload, opening leadership opportunities for teachers, and institutionalizing reforms into the daily life of the department. Third, there are the ongoing conversations that the chair must engage in with the school administrators, such as principals and vice principals, who directly oversee their work in the school governance structure. A supportive administration is a powerful ally in everything from the day-to-day work with teachers to curriculum to change over the long-term. When we say long term, we mean 10 to 15 years. When we say change, we mean both instructional strategies and changes in the way the department conducts itself as a collaborative entity when addressing board and school improvement plans. Remember, departments are both communities and organizations—the reimagined department can create its own dynamic within administrators' plans and then work together with those administrators to provide the direction, support, remediation, and professional development that makes the dynamic a reality. Finally, we look at how you can judge the success of the ongoing work of reimagining the department.

Developing the Collegial Department

We are going to largely ignore the traditional administrative roles of the chair as caretaker, supervisor, inventory taker, equipment and supplies purchaser, timetable developer, exam organizer, and so on. These skills are valuable, but they contribute nothing to leading the creative talents of a staff, demonstrating curriculum leadership, developing a professional learning community (PLC), or steering the faculty toward constructivist thinking as it relates to current instructional and assessment theory. Where they do have importance in reimagining the department is in developing and supporting the moral presence of the chair. Remember, all dealings with teachers need to be based on honesty, courage, care, fairness, and practical wisdom. Here, we will be centering the conversation on the virtues of leadership and the practices of distributed leadership within departments. To illustrate this, we will be using anecdotes from our previous work in these areas.

Virtues and Building the Department

We have reiterated several times that departmental leadership is a moral undertaking. Indeed, that concept underpins much of our writing. In Chapter 3, in considering Peacock's leadership capabilities, we noted that a moral chair is more likely to be effective in reimagining the department as a community committed to improving teaching and learning toward the ideals of the reform documents. Sergiovanni in his work on leadership makes the same point. He first quotes Walton: "Disciplined organizations reflect disciplined leaders whose honed abilities lead them to behave consistently, almost instinctively, in moral ways" (Walton quoted in Sergiovanni 2005, p. 112) and then goes on to indicate himself that "these leaders know and focus on what's important, care deeply about their work, learn from their success, and are trustworthy people" (p. 112).

Further, Sergiovanni argues that, in seeking to develop a collegial environment, leaders need to demonstrate faith, hope, trust, piety, and civility. But what does this mean for a chair who seeks to implement the ideals of the reform documents after he or she has carefully built credibility through science leadership, content knowledge, and working as an advocate for science and science education? Using the example of a department that we have observed over many years may give us some useful insights into transforming these words into concrete actions.

Faith

Sergiovanni (2005) has stated that faith

> *often is communicated as a set of true assumptions. We can hope that once these true assumptions are announced, they will come alive, be accepted, and stir others to action. ... If these assumptions become shared by others in their school community, then a powerful force of ideas will be created. (pp. 114–115)*

But what does this mean for the science chair? In the department that we have worked with extensively, the chair had, since 1988, been researching and trying out science as inquiry in his secondary classes. This involved taking the risk of initiating this learning with his students. As the products (such as research reports and exemplars of classwork) began to accumulate, he made adjustments to refine his approach and the supports he was putting in place for the students. He knew that the strategies were effective and that student attitudes and products were good—better than anything he had done or seen in the previous 10 years. When he became chair in 2000, he began passing student inquiry products around the department, leaving them on desks, and initiating or being drawn into conversations. This start to the discussion about reforming teaching and learning involved the chair building credibility for reform-based teaching strategies:

> *I remember passing around inquiries to the other teachers in the department, and saying, "Look at this work. Isn't it extraordinary?" and leaving it at that. At some point you have to give them credit for picking up and reading the student work, and then thinking, "How could I do that?" A lot of the early discussions were more like what we get when we present material elsewhere now. There was a sort of resistance, but more like "convince me," as opposed to "that won't work" or "I can't do that." From there, the conversations centered on why inquiry was more useful, and how we could get through the curriculum if we were doing inquiry. Eventually, the conversations turned to the question, "If I wanted to try this, then how would I start?"*

The chair had developed a faith in the value of scientific inquiry and the capacity of students to work with it to produce exemplary products. This should ring a bell from Chapter 4: Start in your classroom. Teachers became intrigued, and then the real work began as they first worked to bring themselves up to speed and then to develop their ability to mentor others. This brings us to the virtue of hope.

Hope

Hope is not the same as being wishful. A chair considering where they want to take the department might express a desire to create a working environment in which the ideals of reform documents such as the *Next Generation Science Standards* (*NGSS*; NGSS Lead States 2013) hold primary importance. This is wishful because that is where the effort ends. Such a chair has neither the will nor the plan (goals and actions) to start the department moving in that direction. We've all seen lots of individuals and organizations that think along these lines. Quite often the wish takes hold after attending a professional development session and enthusiasm is high: "Wouldn't it be nice to adopt that or perform like that group?" As time goes on, the wish might recur, but gradually the desire fades as no action is taken to reach for it. The department settles back into mediocrity; it does those things it has always done to get by. For all intents and purposes that is the perception: There goes a department whose faculty works really hard. These are the students we have to work with, this is the curriculum we must deliver, these are our test scores, these are our working conditions, these are our resources, and this is the time available in the day.

Hopefulness then must mean more than being wishful, as Sergiovanni (2005) explains: "Hopeful leaders … react actively to what they hope for and deliberately strive to turn hopefulness into reality." Even if that does not happen right away, "leaders can still be realistic as long as the possibilities for change remain open" (p. 113).

Avoiding wishful thinking means taking deliberate action as well as providing the context for both organizational and individual efficacy. In other words, a chair must be prepared for goal setting, and planning must be ongoing: "Individuals with high hope possess goals, find pathways to these goals, navigate around obstacles, and develop agency to reach their goals" (Sergiovanni 2005, p. 115). Shortly, we will share several vignettes about the work of one science department. As you consider the stories, keep in mind that there was nothing special about this department. It is within a public secondary school (grades 7–12) that serves a broad range of socioeconomic classes, and all the teachers who have worked in the department over the past 15 years would describe themselves as having been educated in the academic tradition of science education. All commenced their own classroom teaching using the strategies that had worked for them in their own education. This is not a criticism of them; it is an acknowledgment of how most of us have been educated. It also means that change is a potential reality for all teachers.

Over a period of three years (2000–2003), and through a combination of ongoing classroom observations, reading, and conversation the teachers began to educate themselves about science as inquiry and then developing ways to transfer that learning into the hands of ninth-grade students. One of the goals that developed was to plan and teach a common introductory unit on scientific inquiry to all students, regardless of their academic ability. This common unit was introduced across the department in 2002 and has been refined since.

The importance of developing this one unit was threefold. By keeping the task manageable, the reforms were seen as realistic. Second, it provided the opportunity to negotiate commonly held meanings for terms such as *inquiry*, *scientific literacy*, and the *scientific method*. As we mentioned before, a lack of definitional clarity is often cited as one of the reasons reforms falter. Only teachers can define for themselves and then meaningfully put into practice the terminology of the reform documents. For the teachers in the following vignettes, it meant embracing a number of instructional and assessment strategies that supported the use of inquiry in the classroom. The instructional strategies include the incorporation of different collaborative learning approaches (e.g., jigsaw, think-pair-share, academic controversy); having students use analogy to show they understand a concept; using conferencing as a small-group instructional strategy; using authentic tasks, rich assessment tasks, and culminating performances to address curricular expectations; and ultimately, using scientific inquiry itself. The assessment strategies include the use of varied tools that do more than simply provide a grade; providing assessment criteria up front; using exemplars to establish how assessment criteria will be applied; developing assessment criteria with students; using exemplars to develop self- and peer-assessment competency with the assessment tool; using exemplars to set the bar and show students what quality work looks like; using assessment formatively to provide for improved performance; and using interviews and conferencing to provide assessment feedback. This commonality also addresses a frequent concern of teachers looking to reform their teaching. Once a common foundation is developed, teachers are free to pursue their reformed-based teaching in their other classes. That means that a student taking chemistry or biology in grade 11 does not have to reinvent the wheel for each teacher or discipline. The only things that change as students move through the senior sciences is the complexity of their writing, the depth and relevance of their background research, their command of data-gathering technology, and their abilities using statistical analysis.

Finally, the development of the common unit led to biology, chemistry, physics, general, and environmental sciences no longer being taught as separate, specialized entities. Through teaching, talking, and modeling the same approaches to inquiry and producing similar assessment products, the teachers came to realize that crosscurricular (subject) knowledge was needed to be successful—just as for researchers in the real world. Students came to see such collegiality on inquiry assessments by their teachers as an important resource for themselves but also as a model for how science is done in the real world. The motivation for this work has been, and continues to be, the observation that students have become more successful at doing and understanding science.

The development of the common grade 9 unit acted as a pathway to other inquiry and constructivist tasks that could be developed to address curricular needs in the units that come later. In terms of assessment, however, the department came to an important decision. There developed a belief that while unit tests were rigorous enough to evaluate knowledge and understanding of conceptual expectations, a culminating performance in which students could demonstrate their knowledge and understanding of scientific inquiry was also needed. The culminating performance took the place of the traditional grade 9 exams at all levels from workplace students to International Baccalaureate students. That innovation occurred in 2003 with the full support of the principal (a topic we shall turn to shortly). In 2004, the department introduced the in-class and in-school science fairs at the 10th-grade level using the same instructional and assessment principles.

Replacing end-of-unit exams with a culminating performance? Heresy! we hear you shout. Remember the courage of convictions we talked about in Chapter 4? This school has won the participation award at the regional science fair for the past 12 years in a row, along with the vast majority of available prizes at the intermediate level. It has also sent a disproportionate share of students to national science competitions. Over the past decade, the science department has also had consistently high numbers of grade 11 and 12 students enrolling in science studies. On average, some 40% of senior students are taking science courses. Anecdotal evidence suggests that these students have been successful in completing postsecondary science-based programs. Needless to say, regardless of their postsecondary destination, students have begun learning about the nature of science and how that knowledge can be put to use—in postsecondary science education, as a problem-solving strategy in their work, or in personal decision making. If it is not clear by now, hopefulness without considered action goes nowhere.

Trust

None of the work and accomplishments produced by this department could have occurred without the development of trust. First, there must be trust by teachers in the assumptions made by the chair that reforming pedagogy will result in positive changes and measurable increases in student success. This is where credibility in the reforms becomes a powerful factor. The teachers must also trust that they will be supported while taking risks to develop the expertise in these reforms and that this support will extend to help when communicating and interacting with people such as parents and administrators. Trust is critical when departments and individuals are attempting a new task. While you must expect bumps in the road as you develop your own initiatives, trust in both what you are doing and in the people in your department. As Sergiovanni (2005) states: "Everyone is vulnerable when trying something new and needs to be assured that mistakes will be accepted and that support will be there" (p. 120). We have seen this time and time again in our work, as evidenced by these two short narratives. The first is by a new teacher commenting on the help afforded her by another teacher when she first started in the department:

> *He gave me resources and different ideas on how to teach the kids the skills that they need for scientific inquiry. He was really helpful at the beginning—he knew I really had no experience with scientific inquiry. I remember him saying, "For the first semester, just start getting comfortable with it, and I'll help you any time. When you're comfortable with it, try two inquiry activities, or if you're more comfortable, maybe try three." What he didn't say was, "No, you have to do it this way, and you have to do five inquiry activities in the first semester." He would always ask me how things were going and give me some really good feedback and suggestions.*

The second narrative is from the chair who acted as the mentor for the chair of the department we are working with:

> *You set up an atmosphere where people feel that they are part of what's going on and that what's going on has importance. They were prepared to try things because it was a safe thing to do, there was no threatening. I trusted them doing it, and they would do it. In a leadership role, you get people to work by creating a suitable environment. You have to create environments where they want to be there, and they want to work together. The two things mesh and you create an environment where people are happy.*

As noted in Chapter 3, creating a trusting environment is critical in the shift from transactional to transitional (and transformative) leadership. The chair's reflection these many years later about this particular shift in the department's direction shows that, without question, the shift was important to the reimagining of the department, and what has been learned on the way has stood the test of time. In the words of the chair, the reforms are no longer "one teacher, it is all of them. It is not one course or grade, it is all of them."

Equally important, the shift has allowed teachers the freedom to experiment and share what they have learned with their colleagues. Even though the scaffolding supporting their work as a department was a common construction, teachers have a unique take on how to tweak the process or "do the science." That sounds like real-world researchers. Following the oral defenses by the ninth graders of their culminating performance reports, a teacher told the chair the following:

> *I never get tired of walking a team through their defense. The learning and insight doesn't stop for them or me even though this is their final evaluation. When I think back to how proud we were of the products, say 10 years ago, and how much better I've become at teaching them about inquiry and facilitating the process ... today's stuff just blows me out of the water.*

Our point here is that the department had evolved into a learning community. Such a community works together to achieve common goals and is characterized "in terms of information gathering, problem solving, the production of creative ideas, and the ability to respond flexibly to new situations or adjust flexibly when interacting with others" (Sergiovanni 2005, pp. 116–117). Or as Yager (2005) writes:

> *Teachers are viewed less as consumers and more as providers of knowledge concerning teaching. Teachers are portrayed less as followers and more as leaders. They are seen less as persons housed in a classroom and more as a member of a professional community. The teacher is not seen as "the target" for change, but as a source and facilitator of change. (p. 18)*

Piety and Civility

The final two virtues that Sergiovanni speaks of are piety and civility—piety in the sense of an uncritical adherence to a particular method. On reading this the first time, the conclusion may be that piety is a destructive force. On rereading the section, it seems clear that Sergiovanni's point is that piety can be tremendously harmful to a department unless members are aware of the danger and work to mitigate against it. One chair we worked with has often said to his department, and his school administration, that the department's "biggest enemy is complacency." He was referring to the danger that teachers might stop evolving after many years of outstanding work and stagnate or slip backward. Teachers and departments can ossify and begin to live in the shadow of past achievements and—despite that stagnation—elevate their status in their own mind's eye:

Piety is a leadership virtue that requires or encourages people to look inward to their own narrow community affiliations. Piety embodies showing loyalty, respect, and affection such as is usually found among friends, comrades in arms, close colleagues at work, and other groups where caring and obligations characterize connections between people. When held together by piety alone, school groups become isolated from one another. Piety demands conformity and justifies exclusion, while civility welcomes diversity, encourages tolerance, and legitimates controversy. Civility builds frameworks within which people can cooperate despite their divergent views and interests. ... Civility draws us outwards to embrace differences. (Sergiovanni 2005, pp. 120–121)

According to Sergiovanni, piety needs a companion, civility, and that is in part what helps keeps departments vital. A civil science department does not close its doors, and in fact does whatever it can to wedge them open. As any chair knows, many factors, often beyond the control of a school, cause school populations to fluctuate. As numbers change, so do staff. Some teachers may opt to voluntarily leave a school while others attempt to transfer in. If a department develops a common purpose, and works with shared assumptions and frameworks around teaching and learning, one might wonder if piety could exclude or devalue any newcomers. One of the initial roles of the chair is to be proactive at offering help and providing support to teachers coming into the department. Over time, this mentoring role can become one that other teachers can take responsibility for.

So, what does it look like for a department to cultivate a culture of civility as a bulwark against piety? One way is to look outside the department for ideas and support. Admittedly, educational research can often come across as impractical and disconnected, but professional journals can provide both the catalyst and substance of ideas to consider and work with. Taking the role of instructional leader seriously may involve cruising through a few research or professional journals and copying, scanning, or emailing articles that hold relevance to certain teachers. Departmental meetings can also be structured around particular articles—the key here is to have teachers read them before the meeting. If some of the readings are relevant to administrators, then copies should be forwarded to them as well.

Another strategy is to work closely with science education faculty within university education programs. Remember, we are all science educators, just in different areas. Reciprocal relationships can see teachers become guest lecturers in their areas of expertise; they can see university faculty working with teachers, students, and administrators on issues of concern to the department or school as well as helping lead professional development sessions for other teachers. One word of advice: Start small and develop the relationship first, just to ensure that everyone is contributing to the promotion of reform-minded teaching and learning. Closely linked to working with university faculty is to work with their students (teacher candidates) on placement. If possible, consider having each teacher candidate work with multiple associates to enhance their experience with inquiry and pedagogy. Make no mistake—this is work for the associates; they are not doing this for money or to find additional time off. True mentoring is hard work.

Finally, be involved in your professional associations. They are only as strong as their membership. Contribute in any way that you and your department can. We all have much to learn from each other. And learning from each other is greatly enhanced by distributing leadership across the department, as we explore next.

Distributing Leadership

In Chapter 3 we cited Axley's 1947 quote about racehorses and plow horses. One strategy for avoiding that fate is to, over time, distribute the leadership functions across the teachers in your department. This does not mean that everyone leads; that is a recipe for disaster. Mayrowetz (2008) described distributed leadership as building human capacity by promoting the notion that as individuals engage in leadership they learn more about themselves and come to a deeper understanding of the issues they face in their work. This development allows departments to address the issues that they face, taking "collective action toward collective goals, internal boundary spanning, and a reliance on expertise, rather than formal authority." Mayrowetz also points out that, unfortunately, "few schools are able to engage in high-level collective inquiry" (p. 431). Our belief is that departments can, over time and with appropriate leadership, develop effective forms of distributed leadership. By appropriate leadership, we mean that the chair must possess formal authority and also intend to exert influence as an instructional leader. The practice of distributed leadership requires leaders who can exercise both formal power and influence. As Harris and Muijs (2005) state:

> *It is clear that certain tasks and functions would have to be retained by those in formal leadership positions but that the key to successful distributed leadership resides in the involvement of teachers in collectively guiding and shaping instructional and institutional development. (p. 34)*

The chair's influence is built by offering credible advice on instructional matters and establishing the trust that allows other teachers to feel confident in stepping forward and beginning to enact their own leadership practices. Remember, this may take three to four years. Continuity in leadership is required to maintain trust. Our work indicates that teachers will need between five and seven years from stepping forward to making substantive contributions to the professional learning of the department and wider science education community. Clearly, these timelines do not align with many board recruitment policies or with those that believe that change can be easily mandated or legislated. These remain significant issues with no clear resolution in sight. Our advice is to do what you can in your particular situation and raise the issue with administrators. And to work most effectively with administrators requires the building of solid reciprocal relationships, as discussed next.

Relationships With School Administrators

The NSTA position statement "Leadership in Science Education" (NSTA 2003) reads,

> *if science leaders and their leadership network are to successfully carry out the roles outlined above, the full support and commitment of the superintendent, the board of education, and the chief state school officer are required. (p. 3)*

This is good, as far as it goes, but it misses the school administrators that teachers have the most contact with. As a chair with a vision to reimagine the department, it is vital that you develop reciprocal long-term relationships with school administrators. Such relationships need to be built on a mutual respect for the roles of each position (see Vignette 4, p. 102). While chairs, as the subject experts, may have a responsibility for instructional leadership in the department, principals have a wider responsibility to both liaise with school boards and to understand different departmental practices and how those practices can contribute to the educational goals of the school. This implies a central role for the principal in the ongoing work of "defining, defending and enabling a viable educational philosophy throughout the school" (Lingard and Christie 2003, p. 329). Hence, a principal would be interested in the following kinds of questions for his or her chairs and departments:

- What does student success look like in your subject?
- What does your subject association regard as best practice?
- How do you know when students are successful?
- What forms of success are transferable between subjects?
- How can the school and department work together to improve student outcomes?

Neither the science department nor the school itself can become a center of harmony without administrators taking a proactive role. For chairs, being able to answer questions such as these from a reform perspective is invaluable because administrators are in a position to offer both tangible and intangible support to teachers and departments. Tangible support includes the provision of resources, time for teacher planning and preparation, public recognition and encouragement, and addressing organization constraints. However, as Supovitz and Turner (2000) warn, tangible supports, while appreciated by teachers, may not be as important in the encouragement of inquiry as the intangible forms of support. They argue that teachers "who felt supported by their principal reported significantly greater use of reform approaches than did teachers who did not feel encouraged by their school leader" (p. 975). Intangible supports can include engagement in the life of the department; managing the politics that invariably occur in schools; addressing parental concerns when they occur; and lobbying authorities, funding agencies, businesses, and universities. A most important form of support, and one that is often overlooked, is the capacity for administrators to facilitate professional learning by increasing the access that teachers have to learning opportunities. More specifically, administrators can encourage teachers to engage in learning that requires them to develop specific school-based solutions to problems of practicing inquiry. Davis (2003) writes:

> *Teachers at all instructional levels must be empowered to create new structures, policies, and practices within their school settings in order to support their collaborations with colleagues and students, the development of goals for change, and their design of and experimentation with innovative instructional and learning practices and assessments. (pp. 25–26)*

Administrators also need to be supportive of their teachers when they learn, share, and critique their professional knowledge with teachers in other schools. Through the expansion of these links and experiences, the knowledge base that teachers need to implement reforms such as the *NGSS* grows. Principals who seek to develop the professional learning capacity of their schools and are prepared to support the professional learning of their teachers will make a significant contribution to the encouragement of reform-based teaching.

The question is, How do we go about initiating reciprocal long-term relationships? The answer lies in developing clear lines of communication with administrators and playing politics (in the best sense of the word). That means involving administrators in the work of the department. Without the support of school administrators, the department will never reach its potential, and the road to reform will be rockier than necessary. Involving administrators in the work of the department does not mean involving them in the minutiae of the day-to-day. It means engaging with administrators in conversations around where the department is and where you would like it to be in the next 5 to 10 years. It means being prepared to argue the case for changes that are planned, and it means being able to show how the department's plans can contribute to the ethos of the school. The example we discussed earlier about the department that developed a common grade 9 science as inquiry unit is a case in point. When the department decided to replace the final grade 9 exam with a culminating performance through which students would demonstrate what they knew and understood about scientific inquiry, the principal was the first person the chair visited to talk to about the initiative. He knew that this could only be successful with administrative support. As should be expected, it was a two-sided conversation. The principal was knowledgeable about curriculum, assessment, and evaluation. He had some concerns about the culminating performance process and how the evaluation would be recorded and reported. Working together, the chair and principal addressed the concerns and by doing so the trust between them continued to build. Equally important, the principal was prepared to back the proposal in the face of any opposition and officially communicated the change to the school community. There was some concern expressed by some parents, but the chair and principal worked together to resolve these issues.

Involving administrators also means making them aware of what is happening in the department. As expertise within the department begins to build (and remember this happens over three to four years minimum), communicate what is happening to both administrators and audiences outside the school building. Opportunities to communicate the work of the department, both successes and challenges, can include organizing reciprocal visits with local teachers to the school, speaking at professional association meetings such as NSTA regional conferences, organizing or leading professional development workshops, submitting articles to professional journals, and working with science educators at local universities.

Keep in mind that some teachers in the department will not want to be involved in such activities. Don't force people. Some teachers may feel that their skills are most useful within the department, and they should be encouraged to work where they feel strongest. Knowing the talents that your teachers possess is an asset that should not be underestimated. Within the department, there is always need for mentoring and working with preservice teachers. Administrative tasks can be delegated. Organizing in-class and school science fairs is an excellent opportunity to invite administrators (and other classes) to attend as well as engage with students in conversations about their learning and products. Our experience is that this becomes a big deal for the students and gives them the opportunity to practice telling peers, administrators, and teachers about their inquiry work. It also becomes a celebration of student achievement that garners good press and positive feelings toward schooling. Both of these are well received by administrators, and by actively developing reciprocal long-term relationships a trust is built that can then be translated into administrative support in terms of budgets, retention of department members during staffing deliberations, release of money to the department to fund supply teachers for leadership activities, and release time to attend conferences and conduct workshops.

One other important area where strong relationships with administrators are crucial is in the hiring of new science teachers. By building an understanding of how the department's reform work helps support the ethos of the school, chairs can work with principals to find teachers with current ability in, *or the potential to develop*, reform-based skills, attitudes, and understandings. Depending on the system used in your board, it may be possible to preview each applicant resume and then interview, rank, and place the successful candidates on a short list. While this may not guarantee that your school rather than another will get the short-listed applicants, it does give the chairs "scouting" knowledge over whom they should look to hire for their available vacancies. Such knowledge is critical in establishing a core group who can become a department of self-regulating learners who in effect become responsible for their own professional learning.

In summary, one of the best ways to keep the administrators in the departmental loop, and onside as supporters, is to invite them to anything of significance the department is doing. If they do not or cannot attend, then take pictures and send them those instead. Closely related to this strategy is to make a point of e-mailing or talking in person with the administrators whenever a teacher or students have done something of significance (such as organize another great in-school science fair, conduct an exemplary inquiry assessment, or complete a waste audit of the school's garbage). Don't take any credit for this, just let administrators get in touch with the teacher and/or students and congratulate them for their efforts. This is a key strategy in building trust within the department and between the department and the administration.

You may be reading this and thinking something like, "This is all well and good, but the administrators in my school are not amenable to anything like this." Our advice, which has been honed by some bitter experiences, is this: You have influence in your department, so work from there. Concentrate on shaping a moral presence in the department and building trust in what you are trying to achieve. This will provide a sense of coherence when conflict arises, which it will.

One of the major sources of conflict is the efforts administrators occasionally make to impose structures on teachers that will align them with the school improvement plan or the (often nebulous) mission statements of the school. The frequent perception is that these impositions are classic top-down management strategies to get a specific product or action that would benefit the top. Relabeled reforms (remember the vice principal from Chapter 2) are also derided as the next fad or the new flavor of the month. If it could just be paid lip service and ignored, then it too would fade away into the ether. Every teacher has his or her own views about all manner of educational theory and practice and, to complicate things even more, personalities and emotions play a role. Some are forceful, some are shy, some are comedians, some are resentful, and some are negative. Some, whether they know it or not, are saboteurs in their actions.

For a chair to deal with these situations requires planning, no small degree of stress, and a conscious decision to concentrate on those projects that can be agreed on to generate some success. In some instances you will need to be tough and remind teachers about the norms that the department operates under, but always follow up conversations and planning with support and treat everyone with respect. If there are disagreements with administrators, keep them private.

If an administrative task needs to be done, time management becomes crucial. Never allow meetings to turn into simple department meetings, always use all of the available time, and try to design the meetings to be collaborative in the sense that members take on work, roles, and responsibilities. You cannot run everything from your "chair" as the boss. Your primary job in these difficult circumstances is to maintain the coherence, collegiality, and direction of the department while working to find a balance between the interests of the department and the mandates of the school. The key to this is to keep teachers' trust in each other and in the work of the department. We keep coming back to trust because it is the linchpin of everything you do. Sergiovanni (2005) writes of a continuum of trust, from those who understand its importance to those who do not:

> *Relational trust was an important catalyst for developing a supportive work culture characterized by school commitment and a positive orientation toward change. It also was an important catalyst for developing a facilitative work structure that included developing a professional community for making decisions together and supporting teacher learning. (p. 119)*

And:

> *Conversely, in schools and school districts that are less effective in bringing about change, trust is an afterthought—often preceded by vision, strategy, and action. Trust gets attention after the school or school district gets into trouble. Leaders typically wind up imposing visions and strategies, which require increased performance monitoring. Resistance usually results, leaving leaders trying to mend fences, improve relationships, and get more people on board. (p. 119)*

As a chair working to improve the teaching and learning that occurs in your department, expect to see both ends of this continuum. Recognize and apply some of the theory you learned about in Chapter 4 when choosing your projects and developing trust, and you too will win through.

In summary, the chair constantly needs to be on the lookout for opportunities that support or increase the department's ability to develop its own reformed professional knowledge base and associated teaching practice. At the same time, you have a responsibility to involve your administration in acquiring, providing opportunity, and utilizing such supports. An administrator may not naturally bring such skill sets to the table, especially early on in his or her careers. This may be something that you need to cultivate over time.

Judging Progress and Success

Standardized test scores do not tell us if our departments are being successful. Because they are firmly rooted in the academic tradition, they offer a narrow definition of success. In this section, we would like to look at factors that might contribute to the overall ethos, or identity, of a reimagined department working toward the ideals of reform documents such as the *NGSS*. In doing this, we are not presenting a checklist to be ticked off, rather we offering "signposts" that may indicate that you and your teachers are actively reforming the teaching and learning that occurs in the department. There are, of course, many avenues available for you to consider what would work best for your department: Very little literature exists about what it means to have success as a secondary science department. What we offer here has been culled from anecdotal evidence provided by chairs we have worked with.

The number of science courses taken by a student over four years in high school varies considerably by school, district, and state. Currently, in Connecticut students must take and pass at least two science courses to graduate, while in Utah students must take and pass three science courses to graduate. While differences in science course graduation requirements vary, there are usually a specific number of required science courses, although it is not mandated that students take science courses each year of their high school career. Given this, one signpost to measure the science department's ability to in attract students is the number of optional courses taken beyond those required for graduation.

Another signpost that might be useful in assessing the impact of a science department is where graduating students are headed. Examples include university, college, apprenticeships, and paid employment. As an example, the second author's school carefully tallies all of those students whose destinations could legitimately be considered to involve further science education and looks at that group as a fraction of the total graduating class. Categories might include nursing, biochemistry, engineering, bioinformatics, medical technology programs, physics, biomaterials, and so on. In the United States, we can imagine a science department tracking their graduates to determine the number of students completing science– technology–, engineering–, and mathematics–focused classes and degrees at the undergraduate level. We recognize that these measures are not easily obtained, but believe departments can stay connected to their students beyond high school, especially given the advances in technologies and social networking sites. And, if this data is measured consistently and methodically, a picture emerges about the relative success of the science department in providing a relevant, science-based, postsecondary future for the students. In the second author's school where this measure has been used, the percentage of students going on with science in some form has remained between 25 and 35 percent of the graduating class over the past decade.

As a high school teacher, you may be aware that many schools measure their vibrancy through student engagement in their intramural and extramural (varsity) sports programs, student clubs, student council, yearbook and so on. That same measure could be another signpost for your department: How does your department provide opportunities that enrich science students' sense of belonging, accomplishment, interest, or success or help students build science identities? For example, are there science clubs; science tutorial opportunities; a public speaker series; in-class guest speakers; or university, college, and community-based science camps? Do your students have the opportunity to compete at an appropriate level? Examples include varsity Envirothon (*www.envirothon.org*) teams, Science Olympiad (*http://soinc.org*) teams, and academic contests such as the University of Toronto National Biology competition (*www.biocomp.utoronto.ca*; many U.S. schools compete in this one), or the USA Science and Engineering Festival (*www.usasciencefestival.org*). What about in-class, school, regional, and national science fairs? Make sure you look at the NSTA position statement"Science Competitions."

We talked earlier about professional reading and professional development, and there are signposts here as well. Many teachers already keep their finger on the pulse of local, state, and national trends as well as research and events by belonging to professional associations and subscribing to their online and/or print journals and newsletters (e.g., *Georgia Science Teachers Association Newsletter*). Can you and your department share your experiences with your colleagues in other schools? The professional organizations we belong to also offer regional and national conferences, and attendance at these can be an indication of engagement and vitality. Perhaps members of your staff could be encouraged to engage in the production or facilitation of sessions and workshops that highlight your work? The fabled journey of a thousand miles starts with a single step.

Final Comments

The department may be viewed as a fertile field on which successive years of student crops can be grown. While many students will "find their way" during this time and select postsecondary destinations in science, our concern with reimagining the department should be to promote science education and find ways to make it relevant and useful to all our students. Regardless of sex, race, or socioeconomic status, we want to see students value science as a viable and potentially enriching part of their lives.

To reimagine the department, it is critically important that chairs understand that

> *A strong heartbeat is a [department's] best defense against the obstacles leaders face as they work to change [departments] for the better. Strengthening the heartbeat of [departments] requires that we rethink what is leadership, how leadership works, what is leadership's relationship to learning, and why we need to practice leadership and learning together. (Sergiovanni 2005, p. 122)*

Building trust within the department is paramount. Faith in our colleagues and the assumptions that reimagining the department is based on emanates from trust. If we attempt to plow ahead, unprepared (or unwilling) to develop a moral presence, then we are setting ourselves up for failure. It's a self-perpetuating cycle. Low trust means an isolated staff, and the longer this goes on the lower the level of trust sinks. Teachers become withdrawn from the department as community, keep good ideas to themselves, and the cycle spirals lower and lower. But it doesn't need to be like that.

We wish you, and your department, the very best in the challenge of reimagining and then realizing your vision of what a science department can be. Leading a paradigm shift in thinking and practice of any magnitude is a challenge that requires leadership based on hope, trust, faith, and civility from both the chair and the department, supported by school administrators.

Summary

- Departments do not work in isolation from the rest of the school. To reimagine the department is to also be active in developing strong political and practical relationships with school administrators. The aim of reimagining the department is to develop a long-term culture that is simultaneously owned by the teachers and supported by school administrators.
- A supportive administration is a powerful ally not only in terms of your daily, weekly, monthly, semester, and school-year routines, but also

change over the long term. The reimagined department can create its own dynamic within administrators' plans; and then work together with those administrators to provide direction, support, remediation, and professional development that makes the dynamic a reality.

- All dealings with teachers need to be based on honesty, courage, care, fairness and practical wisdom. While we acknowledge the moral aspects of these administrative roles, reimagining the department is centered on the virtues of leadership, and the practices of distributed leadership within departments.
- A chair must be prepared for goal setting, and planning must be ongoing.
- Trust is critical when departments and individuals are attempting a new task. While you must expect bumps in the road as you develop your own initiatives, trust in both what you are doing and the people in your department.
- Distribute the leadership functions across the teachers in your department. The chairs' influence is built by offering credible advice on instructional matters and establishing the trust that allows other teachers to feel confident in stepping forward and beginning to enact their own leadership practices.
- As a chair with a vision to reimagine the department, it is vital that you develop reciprocal long-term relationships with school administrators.
- To initiate reciprocal long-term relationships with administrators, develop clear lines of communication with administrators. This means involving them in the work of the department. Involving administrators in the work of the department does not mean involving them in the minutiae of the day-to-day. It means engaging with administrators in conversations around where the department is, and where you would like it to be in the next 5 to 10 years. It means being prepared to argue the case for changes that are planned, and it means being able to show how the department's plans can contribute to the ethos of the school.

Vignette 4

JEFF UPTON

Mr. Jeff Upton has been a secondary vice principal for 14 years with Lakehead Public Schools in Ontario, Canada. For six years he was the vice principal in the school where Doug Jones is the chair. He has a masters of education from the University of Toronto and has recently taken on the role of education officer in the local board office. He is also past-President of Lakehead District Ontario Principals Council. We have worked with Jeff for a decade and asked him to comment on the role of school administrators in supporting departmental reforms of teaching and learning.

Building political support in schools is not an area that is often spoken about, let alone taught. Though recent research seems to be delving into this area, most administrators and chairs wade into this milieu knowingly, yet often without a clear plan or the skills necessary to achieve their goals and intent.

In my observations, school politics is a result of competition for scarce resources, attempts to influence policy, implementation of programs, competing aspirations, and various views about how to achieve goals that are set out for schools. The micropolitics by which schools are shaped occur through the disputes, conflicts, and problems in which individuals often focus upon strategies, influence, and knowledge to achieve their goals rather than using rules, power, and status. Administrators and chairs can also come into conflict over competing visions and the level of professional collaboration needed for achieving goals. Micropolitical tensions can occur when administrators are trying to encourage chairs to accomplish tasks for student success that involve changing practices, especially when these are not communicated clearly or developed together in a collaborative manner. If a department is comfortable in its situation, to then be informed by administrators that change is necessary, they will feel the tension—especially when they don't understand the reason and purpose of the change. This situation, which is too frequent, comes down to the issues of communication and relationships. As an administrator you must be a good communicator; you must be clear about the purpose of initiatives that are brought forward and the reasons for the change. If departments understand what the purpose of a change is, and they have been part of the process to identify reforms and understand that the purpose is specific to improving student success, they are in a better position to work with administrators. Communication, empathy, and synergy are the skills that develop the relationships between administrators and chairs, and it is those professional relationships that help move student success initiatives forward.

Given the importance of relationships, there is a developmental need for both the professional and personal relationships between administrators and chairs to be built on trust. Strategically, the development of trust will allow leaders to take risks in ways that support student success. I have found over the years that the key to implementing and sustaining initiatives for long-term success is the quality of the relationships developed between administrators and chairs.

Two important points need to be made about developing these relationships.

First, department leadership really is developed when synergy and distributed leadership practices (i.e., the distribution of influence among leaders that shapes a school's culture) are put into place. Distributed leadership practices, based on trusting relationships, allow chairs to exercise their formal positional power along with their relational influence on teachers, knowing they are supported by administration. Second, let's not fool ourselves: relationships take time to develop. It is important for relationships and trust to develop by giving people the opportunity to work together for an extended period. This long-term continuity is so important because, as people get to know each other, there is the development of the deep trust that leads to a deeper professional relationship. It is this deep relationship that allows chairs, supported by administrators, to lead changes in teaching and learning by taking intentional and deliberate actions that will lead us to sustainable student success. As an aside, there will still be arguments and tension within trusting relationships, so remember to continue nurturing them.

The importance of building trust and political support cannot be understated. The issue of trust is a precondition to the creation of distributed leadership and synergy within the school. The trust relationship between administrators and chairs should be seen as a foundational part of creating conditions for school success and student improvement. Developing a synergistic trust environment where a distributed leadership environment is created and supported and where chairs are afforded the opportunity to make decisions about curriculum and assessment based on intentional and deliberate actions leads to an environment in which teachers are able to fully act as the educational professionals they are. When teachers are afforded trust and political support by administrators, it empowers them to seek out best practices for students. In a trusting environment, teachers are able to take professional risks in challenging current instructional practices, knowing that they will be supported by their administrators.

The ultimate goal is to create a micropolitical environment that gives departments opportunity to reform teaching and learning to improve student learning and success, while at the same time developing an atmosphere of distributed leadership in which ownership of each student's success occurs. Though micropolitics and distributed leadership are two entities unto themselves, they go hand in hand in a school so that administrators, chairs, and teachers can achieve the best learning conditions for students and, ultimately, student success. Administrators need to be proactive in building relationships and political support by developing the trust that allows for open and honest dialogue and conversation. Only by having these conversations can chairs and their teachers truly have the freedom and capacity to question their beliefs, actions, and practices. When this is the case, and individuals do not feel threatened, changes to teaching and learning in line with subject-specific best practices can lead to real and sustained improvements in student success.

Where Am I Today? Questions to Ask Yourself (and Some Ideas to Ponder)

For You as a Department Chair

1. What are you "hopeful" about concerning your department, its performance, and its potential direction or development? What is this hope based on? history, emotion, experience?
2. Consider the voices for reform (*NGSS* and instructional and assessment strategies related to it) and have short but purposeful conversations with individual teachers about their hopes.
3. Sit down with your principal and have an honest conversation about what his or her thoughts are about where the department is and what might be possible (think trust and committing support down the line).
4. Make a list, prioritize, and create an action plan. Think about what's in reach, what might be successful immediately, and what might need more time to lay the groundwork for your teachers as well as the supports necessary to be successful (don't be afraid to use categories such as upcoming year, one to three years, long term; don't ignore the last two categories). You need to become purposeful about them or they will always remain that distance away from you.
5. What will you do to support your staff to develop in their professional learning?
6. How can their individual learning lend support to your departmental reform?
7. Consider having your teachers complete an annual learning plan containing their hopes and action plans for the coming instructional year. (If your school does this already, encourage your staff to share their thoughts with you on an individual basis.) Don't just file these away. Think: Where might my support make a difference for them?
8. How can you focus your teachers on paying attention to their words or plans? One way would be to conduct interviews at the end of the term with the growth plans in hand and ask questions. What has been successful? What should be maintained and nurtured? What needs a different plan or more support to be successful?
9. Another strategy might be to be to investigate the PLC structure and operational processes. Perhaps learning about this could be a goal of your own. This might be a longer-term plan, however, simply because finding the time to bring a department together willingly to start this process can be contentious. You might need to develop board and school administrative support first. Have you thought about going to a PLC conference as a start? Do the science chairs of the board meet as a group? If not, how would you propose this look, and what would the group's goals or discussions be about?

10. Think about the terms *intramural* and *extramural* for a minute. They are terms that apply to you and your department.
11. What does your department do besides scheduled lessons to support the learning, growth, and success of students?
12. What does your department do in a "science" sense to goes beyond the boundaries of your classrooms?
13. What do you need to do to prioritize, plan, build support from the appropriate teachers and administrators, develop trust, and think about the "right time"?
14. Have you really considered trust as a key virtue and essential component of your leadership?
15. What concrete things can you do to develop trust in both yourself and in the teachers in your department?
16. Who are the leaders and administrators within the school, and board, where trust will build bridges and support down the line?
17. Think about your network outside of the department for a minute. These are the people and organizations that might be able to support your work in any number of ways, be it financial, curricular, professional development, public relations, political, department profile boosters, equipment, supplies, and so on.
18. Make a list; make a plan; make an informal contact; trade news, thoughts, and float initiatives you might be able to help each other with; develop trust. Reflect on the conversation and make notes; think about ways to keep this group in your region of influence.

REFERENCES

Aikenhead, G. 2006. *Science education for everyday life.* New York: Teachers College Press.

American Association for the Advancement of Science (AAAS). 1993. *Benchmarks for science literacy.* Washington, DC: AAAS.

Axley, L. 1947. Head of department: A racehorse with plowhorse duties. *The Clearing House* 21 (2): 274–276.

Bain, K. 2004. *What do the best college teachers do?* Cambridge, MA: Harvard University Press.

Baker, W. P., M. Lang, and A. E. Lawson. 2002. Classroom management for successful student inquiry. *Classroom Management* 75 (5): 248–252.

Bell, P., L. Bricker, C. Tzou, T. Lee., and K. Van Horne. 2012. Exploring the science framework: Engaging learners in scientific practices related to obtaining, evaluating, and communicating information. *Science Scope* 36 (3): 17–22.

Berland, L. K. 2011. Explaining variation in how classroom communities adapt the practice of scientific argumentation. *Journal of the Learning Sciences* 20 (4): 625–664.

Blenkin, G. M., G. Edwards, and A. V. Kelly. 1997. Perspectives on educational change. In *Organizational effectiveness and improvement in education*, ed. A. Harris, N. Bennett, and M. Preedy, 216–230. Buckingham, UK: Open University Press.

Bourdieu, P. 1984. *Distinction: A social critique of the judgement of taste.* Cambridge, MA: Harvard University Press.

Bourdieu, P. 1990. *The logic of practice,* trans. R. Nice. Stanford, CA: Stanford University Press.

Brock, W. H. 1975. From Liebig to Nuffield. A bibliography of the history of science education, 1839–1974. *Studies in Science Education* 2 (1): 67–99.

Brundrett, M., and I. Terrell. 2004. *Learning to lead in the secondary school: Becoming an effective head of department.* London: RoutledgeFalmer.

Busher, H., and A. Harris. 1999. Leadership of school subject areas: Tensions and dimensions of managing in the middle. *School Leadership and Management* 19 (3): 305–317.

Bybee, R. W. 1993. *Reforming science education: Social perspectives and personal reflections.* New York: Teachers College Press.

Bybee, R. 2011. Scientific and engineering practices in K–12 classrooms: Understanding a framework for K–12 science education. *The Science Teacher* 78 (9): 34–40.

Calabrese Barton, A., and K. Yang. 2000. The case of Miguel and the culture of power in science. *Journal of Research in Science Teaching* 37 (8): 871–889.

Campbell, T., P. S. Oh, and D. Neilson. 2012. Discursive modes and their pedagogical functions in model-based inquiry (MBI) classrooms. *International Journal of Science Education* 34 (15): 2393–2419.

Carlone, H. B. 2003. Innovative science within and against a culture of 'achievement.' *Science Education* 87 (3): 307–328.

Clandinin, D. J., and F. M. Connelly. 1995. *Teachers' professional knowledge landscapes.* New York: Teachers College Press.

Darling-Hammond, L., and G. Sykes. 1999. *Teaching as the learning profession: Handbook of policy and practice.* San Francisco, CA: Jossey-Bass.

Davis, K. S. 2003. "Change is hard": What science teachers are telling us about reform and teacher learning of innovative practices. *Science Education* 87 (1): 3–30.

Dewey, J. 1916. *Democracy and education.* New York: The Free Press.

Duschl, R., H. Schweingruber, and A. Shouse, eds. 2007. *Taking science to school: Learning and teaching science in grades K–8.* Washington, DC: National Academies Press.

Eggleston, J. 1977. *The sociology of the school curriculum.* London: Routledge and Kegan Paul.

Fyfe, W. H. 1934. Science in secondary education. *The School* 22: 653–660.

Goodson, I. F. 1993. *School subjects and curriculum change.* 3rd ed. London: Falmer Press.

Goodson, I. F., and C. J. Marsh. 1996. *Studying school subjects.* London: Falmer Press.

Hall, G. E., and S. M. Hord. 2011. *Implementing change: Patterns, principles and potholes.* 3rd ed. Upper Saddle River, NJ: Pearson.

Hargreaves, D. H. 2000. The knowledge creating school. In *Leading professional development in education*, ed. B. Moon, J. Butcher, and E. Bird, 224–240. London: RoutledgeFalmer.

Harris, K., F. Jensz, and G. Baldwin. 2005. *Who's teaching science? Meeting the demand for qualified science teachers in Australian secondary schools.* Melbourne, Australia: Centre for the Study of Higher Education. *www.cshe.unimelb.edu.au/people/harris_docs/Who'sTeachingScience.pdf*

Harris, A., and D. Muijs. 2005. *Improving schools through teacher leadership.* Maidenhead, UK: Open University Press.

High Level Group on Human Resources for Science and Technology. 2004. *Europe needs more scientists.* Brussels: European Commission. *http://ec.europa.eu/research/press/2004/pr0204en.cfm*

Hodson, D. 1988. Science curriculum change in Victorian England: A case study of the Science of Common Things. In *International perspectives in curriculum history,* ed. I. F. Goodson, 139–178. Beckenham, UK: Croom Helm.

Hodson, D. 2003. Time for action: Science education for an alternative future. *International Journal of Science Education* 25 (6): 645–670.

Hodson, D. 2011. *Looking towards the future.* Rotterdam, the Netherlands: Sense Publishers.

Hord, S. M. 1997. Professional learning communities: What are they and why are they important? *Issues … About Change* 6 (1). *www.sedl.org/pubs/catalog/items/cha35.html*

Jackson, P. W. 1986. *The practice of teaching.* New York: Teachers College Press.

Kliebard, H. M. 1986. *The struggle for the American curriculum, 1893–1958.* Boston: Routledge and Kegan Paul.

Krajcik, J., and J. Merritt. 2012. Engaging students in scientific practices: What does constructing and revising models look like in the science classroom? Understanding a framework for K–12 science education. *Science Scope* 35 (7): 6–10.

Lampert, M. 1985. How do teachers manage to teach? Perspectives on problems of practice. *Harvard Education Review* 55: 178–194.

Layton, D. 1973. *Science for the people: The origins of the school science curriculum in England.* London: George Allen and Unwin.

Layton, D. 1981. The schooling of science in England, 1854–1939. In *The parliament of science,* ed. R. MacLeod and P. Collins,188–210. Northwood, UK: Science Reviews Ltd.

Lederman, N. G. 2007. Nature of science: Past, present, and future. In *Handbook of Research on Science Education*, ed. S. K. Abell and N. G. Lederman, 831–880. Mahwah, NJ: Lawrence Erlbaum and Associates.

Lingard, B., and P. Christie. 2003. Leading theory: Bourdieu and the field of educational leadership. An introduction and overview to this special issue. *International Journal of Leadership in Education* 6 (4): 317–333.

Lord, B. 1994. Teachers' professional development: Critical colleagueship and the role of professional communities. In *The future of education: Perspectives on national standards in education in America*, ed.

N. Cobb, 175–204. New York: College Entrance Examination Board.

Mayes, R., and T. R. Koballa. 2012. Exploring the science framework. *The Science Teacher* 79 (9): 8–15.

Mayrowetz, D. 2008. Making sense of distributed leadership: Exploring the multiple usages of the concept in the field. *Educational Administration Quarterly* 44 (3): 424–435.

Melville, W., and A. Bartley. 2010. Mentoring and community: Inquiry as stance and science as inquiry. *International Journal of Science Education* 32 (6): 807–828.

Metz, M. H. 1986. *Different by design: The context and character of three magnet schools.* New York: Routledge and Kegan Paul.

Michaels, S., A. Shouse, and H. Schweingruber. 2008. *Ready, set, science!* Washington, DC: National Academies Press.

National Research Council (NRC). 1996. *National Science Education Standards.* Washington, DC: National Academies Press.

National Research Council (NRC). 2000. *Inquiry and the national science education standards: A guide for teaching and learning*. Washington, DC: National Academies Press.

National Research Council (NRC). 2005. *America's lab report: Investigations in high school science.* Washington, DC: National Academies Press.

National Research Council (NRC). 2012. *A framework for K–12 science education: Practices, crosscutting concepts, and core ideas*. Washington, DC: National Academies Press.

National Science Teachers Association (NSTA). 2003. Leadership in science education. *www.nsta.org/about/positions/leadership.aspx*

National Science Teachers Association (NSTA). 2003. Science competitions. *www.nsta.org/about/positions/competitions.aspx*

National Science Teachers Association (NSTA). 2006. Professional development in science education. *www.nsta.org/about/positions/profdev.aspx*

National Science Teachers Association (NSTA). 2007. The integral role of laboratory investigations in science instruction. *www.nsta.org/about/positions/laboratory.aspx*

National Science Teachers Association (NSTA). 2011. Quality science education and 21st-century skills. *www.nsta.org/about/positions/21stcentury.aspx*

National Science Teachers Association (NSTA) 2012. Learning science in informal environments. *www.nsta.org/about/positions/informal.aspx*

National Science Teachers Association (NSTA). 2013. The *Next Generation Science Standards. www.nsta.org/about/positions/ngss.aspx*

NGSS Lead States. 2013. *Next Generation Science Standards: For states, by states.* Washington, DC: National Academies Press. *www.nextgenscience.org/next-generation-science-standards.*

NGSS Lead States. 2014. Educators Evaluating the Quality of Instructional Products (EQuIP) Rubric for Lessons and Units: Science. *www.nextgenscience.org/sites/ngss/files/EQuIP%20Rubric%20for%20Science%20October%202014_0.pdf*

O'Sullivan, C. Y., and A. R. Weiss. 1999. *Student work and teacher practices in science.* U.S. Department of Education, Washington, DC: Office of Educational Research and Improvement, National Center for Education Statistics.

Peacock, J. S. 2013. Making your own role: Exploring instructional leadership practice of high school science department chairs. PhD diss., University of Georgia, Athens.

Peacock, J. S. 2014. Science instructional leadership: The role of the department chair. *Science Educator* 23 (1): 36–48.

Public Schools Commission. 1864. Evidence of Michael Faraday, 18 November 1862. *Parliamentary Papers* 4 (2): 375–382.

Reid, W. A. 1985. Curriculum change and the evolution of educational constituencies: The English sixth form in the nineteenth century. In *Social histories of the secondary curriculum: Subjects for study,* ed. I. F. Goodson. London: Falmer Press.

Rinker, F. 1950. The department head. *NASSP Bulletin* 34 (174): 48–53.

Robinson, V. M. J. 2010. From instructional leadership to leadership capabilities: Empirical findings and methodological challenges. *Leadership and Policy in Schools* 9 (1): 1–26.

Rudolph, J. L. 2005. Turning science to account: Chicago and the general science movement in secondary education, 1905–1920. *Isis* 96 (3): 353–389.

Schwab, J. J. 1958. The teaching of science as inquiry. *Bulletin of the Atomic Scientists* 14 (9): 374–379.

Schwab, J. J. 1962. The teaching of science as enquiry. In *The teaching of science,* ed. J. Schwab and P. Brandwein. Cambridge, MA: Harvard University Press.

Sergiovanni, T. J. 1992. *Moral leadership: Getting to the heart of school improvement*. San Francisco, CA: Jossey-Bass Inc.

Sergiovanni, T. J. 2005. The virtues of leadership. *The Educational Forum* 69: 112–123.

Sheppard, K., and D. M. Robbins. 2007. High school biology today: What the Committee of Ten actually said. *Life Sciences Education* 6 (3): 198–202.

Silins, H. C. 1994. The relationship between transformational and transactional leadership and school improvement. *School Effectiveness and School Improvement* 5 (3): 272–98.

Simonton, D. K. 2013. After Einstein: Scientific genius is extinct. *Nature* 493 (7434): 602

Siskin, L. S. 1994. *Realms of knowledge: Academic departments in secondary schools.* London: Falmer Press.

Starratt, R. J. 1999. Moral dimensions of leadership. In *The values of educational administration,* ed. P. T. Begley and P. E. Leonard, 22–35. London: RoutledgeFalmer.

Stigler, J. W., and J. Hiebert. 1999. *The teaching gap: Best ideas from the world's teachers for improving education in the classroom*. New York: The Free Press.

Supovitz, J. A., and H. M. Turner. 2000. The effects of professional development on science teaching practices and classroom culture. *Journal of Research in Science Teaching* 37 (9): 963–980.

Teacher Training Agency (TTA). 1998. *National standards for subject leaders.* London: Her Majesty's Stationery Office.

Tomkins, G. S. 1986. *A common countenance: Stability and change in the Canadian curriculum.* Scarborough, ON: Prentice Hall.

Tytler, R. 2007. *Re-imagining science education: Engaging students in science for Australia's future.* Camberwell, Victoria, Australia: Australian Council for Educational Research.

van Driel, J. H., and N. Verloop. 1999. Teachers' knowledge of models and modelling in science. *International Journal of Science Education* 21 (11): 1141–1153.

van Langenhove, L., and R. Harré. 1999. Introducing positioning theory. In *Positioning theory: Moral contexts of intentional action,* ed. R. Harré, and L. van Langenhove, 14–31. Malden, MA: Blackwell Publishers.

Weller, L. D. 2001. Department heads: The most underutilized leadership position. *NASSP Bulletin* 85 (622): 73–81.

White, R. T. 1988. *Learning science.* Oxford, UK: Basil Blackwell.

Wildy, H., and J. Wallace. 2004. Science as content: Science as context: Working in the science department. *Educational Studies* 30 (2): 99–112.

Yager, R. E. 2005. Achieving the staff development model advocated in the national standards. *Science Educator* 14 (1): 16–24.

INDEX

Page numbers printed in **boldface** *type refer to figures or tables.*

S